Painting, Science, and the Perception of Coloured Shadows

Many artists and scientists – including Buffon, Goethe, and Philipp Otto Runge – who observed the vividly coloured shadows that appear outdoors around dawn and dusk, or indoors when a candle burns under waning daylight, chose to describe their colours as 'beautiful'.

Paul Smith explains what makes these ephemeral effects worthy of such appreciation – or how depictions of coloured shadows have genuine aesthetic and epistemological significance. This multidisciplinary book synthesises methodologies drawn from art history (close pictorial analysis), psychology and neuroscience (theories of colour constancy), history of science (the changing paradigms used to explain coloured shadows), and philosophy (theories of perception and aesthetic value drawn from Wittgenstein and Merleau-Ponty).

This title will be of interest to scholars in art history, art theory, and the history of science and technology.

Paul Smith is Professor of History of Art at the University of Warwick.

Cover image: Carl Gustav Carus, *A River Cruise on the Elbe River Near Dresden (A Morning on the Elbe River)*, 1827 (detail). Oil on canvas, 29 × 22 cm. Düsseldorf, Kunstpalast. Photo: Kunstpalast/Paul Smith.

Science and the Arts since 1750

Series Editors:

Barbara Larson

University of West Florida

Ellen K. Levy

Independent Artist and Writer

This series of monographs and edited volumes explores the arts – painting and sculpture, drama, dance, architecture, design, photography, popular culture materials – as they intersect with emergent scientific theories, agendas, and technologies, from any geographical area from 1750 to now.

Photography, Natural History and the Nineteenth-Century Museum
Exchanging Views of Empire
Kathleen Davidson

Art, Technology and Nature
Renaissance to Postmodernity
Edited by Camilla Skovbjerg Paldam and Jacob Wamberg

The Organic School of the Russian Avant-Garde
Nature's Creative Principles
Isabel Wünsche

Science, Technology, and Utopias
Women Artists and Cold War America
Christine Filippone

Visualizing the Body in Art, Anatomy, and Medicine since 1800
Models and Modeling
Edited by Andrew Graciano

Painting, Science, and the Perception of Coloured Shadows
'The Most Beautiful Blue'
Paul Smith

For more information about this series, please visit: www.routledge.com/Science-and-the-Arts-since-1750/book-series/ASHSER4039

Painting, Science, and the Perception of Coloured Shadows

'The Most Beautiful Blue'

Paul Smith

Routledge
Taylor & Francis Group
NEW YORK AND LONDON

First published 2021
by Routledge
52 Vanderbilt Avenue, New York, NY 10017

and by Routledge
2 Park Square, Milton Park, Abingdon, Oxon, OX14 4RN

Routledge is an imprint of the Taylor & Francis Group, an informa business

Library of Congress Cataloging-in-Publication Data
Names: Smith, Paul, 1956- author.
Title: Painting, science, and the perception of coloured shadows: 'the most beautiful blue'/Paul Smith.
Description: New York: Routledge, 2021. | Includes bibliographical references and index.
Identifiers: LCCN 2020044068 (print) | LCCN 2020044068 (ebook) | ISBN 9781138488090 (hardback) | ISBN 9781351042024 (ebook)
Subjects: LCSH: Color (Philosophy) | Shades and shadows. | Painting – Philosophy. | Shades and shadows in art.
Classification: LCC B105.C455 S65 2021 (print) | LCC B105.C455 (ebook) | DDC 752—dc23
LC record available at https://lccn.loc.gov/2020044068
LC ebook record available at https://lccn.loc.gov/2020044069

ISBN: 978-1-138-48809-0 (hbk)
ISBN: 978-1-351-04202-4 (ebk)

Typeset in Sabon
by Apex CoVantage, LLC

For Miranda

de vastes ombres bleues chassent en cadence devant elles la foule des tons orangés et rose tendre qui sont comme l'écho lointain et affaibli de la lumière.

Charles Baudelaire, *Salon de 1846*

Contents

Illustrations

Acknowledgements

If this book has an origin, it was the experience of coloured shadows I had in 1982 that I describe in the Introduction. It was only much later, however, that I began to collect my thoughts about them, when Peter Mack kindly asked me to speak at the conference on colour held at the Warburg Institute in June 2014. It subsequently received a shot in the arm when Sonia Sedivy generously invited me to present a paper at the colloquium on art and perception she hosted in Toronto in May 2017. I am grateful to all those present, but especially to Bence Nanay and Dominic Lopes for their supportive response to my paper, which gave me the encouragement to submit my manuscript to Routledge later that month. The following March, I was equally heartened when Claude Imbert told me she enjoyed the paper on 'ombres colorées' I presented at the SERD conference in Paris, *L'Oeil du XIXe siècle*.

Not long afterwards, Sonia Sedivy and John Hyman were good enough to include a version of my Toronto paper in the special issue of the *British Journal of Aesthetics* that appeared in October 2018. And Érika Wicky and her colleagues at SERD saw fit to publish a version of my Paris paper in the *actes du colloque* that went online in December 2019.

Writing this book was particularly difficult for me as an art historian because it involved acquiring a basic understanding of recent psychological and neuroscientific theories of colour constancy. I am especially grateful therefore to Anya Hurlbert and David Foster for their infinitely patient and unwaveringly constructive responses to my many, often hopelessly naive questions. I must also thank John Mollon, whose work on Gaspard Monge has been foundational to my project, for alerting me to the pitfalls awaiting the unwary trespasser into this field.

My primary scholarly debt, however, is to the late Michael Baxandall for the handful of brilliant pages he wrote on the historical understanding of coloured shadows. I have profited in addition from the deeply serious research into coloured shadows undertaken by Georges Roque. I must also acknowledge a debt to Francesca Fiorani's recent work on Leonardo da Vinci's coloured shadows and to Janis Bell for her writing on Leonardo's theory of colour more generally. More recently, I have benefitted from Julia Schöller's richly descriptive and lavishly illustrated book on coloured shadows in later nineteenth-century art. Of the older literature on the subject, Oscar Reutersvärd's essay of 1950 on 'The *Violettomania* of the Impressionists' remains a touchstone, as does Alan Staley's evergreen study of the Pre-Raphaelites, first published in 1973.

All translations in the text are my own, unless otherwise stated. Among those I have borrowed are the translations of Leonardo's *Trattato della pittura* made by Claire

Farago and Janis Bell for *The Fabrication of Leonardo da Vinci's Trattato della pittura*, published by Brill in 2018. I have also taken excerpts from David Britt's translations of Carl Gustav Carus's writings, which appeared in *Nine Letters on Landscape Painting* published by the Getty Research Institute in 2002. All translations from German sources unavailable in English were made by Elena Theodorakopoulos.

I am greatly indebted to Stephanie Nebbia for granting me access to the George Field archives housed at Winsor & Newton. I must also thank Giorgia Bottinelli for allowing me to peruse the Joseph Clover papers kept at Norwich Castle Museum. Every effort has been made to trace copyright holders and obtain relevant permissions for the passages quoted from these and other archival sources.

A number of curators were kind enough to allow me to inspect and photograph paintings which I discuss in this book. In the UK, these include Colin Harrison at the Ashmolean Museum, Carol Jacobi at Tate Britain, Robert Wenley at the Barber Institute, and Hannah Williamson at Manchester Art Gallery. Among those in Germany who helped me are Hanna Bahr at the Kunstpalast in Düsseldorf, Markus Bertsch at the Hamburger Kunsthalle, and Anna Marie Pfäfflin at the Berlin Kupferstichkabinett.

The Clark Institute awarded me a summer fellowship in 2015. I am much indebted to those responsible for this vital period of study and reflection, which enabled me to make considerable headway with this project. I should also acknowledge my especial thanks to the Getty Research Institute, where I spent a wonderful year in 2007–08, reading colour science. It was here that I also experienced the Damascene moment involving the green shadow described in the Introduction.

I am lucky enough to have had colleagues at Warwick who shared my enthusiasm for coloured shadows, particularly Lorenzo Pericolo, who invited me to speak to his students about them. I must also thank Michael Hatt for allowing me to join him on his excursions to Denmark in pursuit of Eckersberg and for joining me on my trip to Düsseldorf to see Carus.

I am grateful to Isabella Vitti at Routledge for placing her faith in this project and to Katie Armstrong for answering my interminable questions with consummate professionalism and good humour. I am equally indebted to Barbara Larson, the series editor of Science and the Arts since 1750, for her many insightful comments on my manuscript.

My greatest debt is nevertheless to my wife, Elena. I must thank her, more especially, for the unstinting support she gave me for the whole of this book's long gestation period. Without her, writing it would have been impossible.

Introduction

It may seem eccentric to devote an entire book to phenomena as transient and insubstantial as the particular kind of coloured shadows generated by the interplay of two differently coloured light sources, as when the warm sun sets or rises in a clear sky or a candle burns under waning daylight or – more rarely – in bright sunshine under a deep blue sky. Notwithstanding, it came about because I had two indelible experiences in which they appeared so extraordinarily beautiful that I felt compelled to understand them better. The first occurred one gloriously sunny July afternoon in 1982 on a ferry boat just outside Le Havre. As I went up on deck, I was forcibly struck by the splendour of the blue shadows on the boat's white prow, which were just as saturated as the French blue boiler suit of the engineer leaning over its edge. The photograph I took on a slide film has since deteriorated so much as to be unserviceable. So it was with some relief that I encountered an equally intense blue shadow just after 1.30 p.m., one March afternoon in 2018, in the foothills of the Mont Sainte-Victoire (Figure 0.1) near Le Tholonet in Provence. My second experience occurred at dusk on the plaza of the Getty Research Institute in Los Angeles in December 2007. Although the sun was already sinking into the Pacific Ocean, its last rays could still reach the travertine marble walls of this elevated building, giving them a spectacular, almost incandescent warmth. At the same time, I noticed, the shadow cast by my daughter, Miranda, was a deep but luminous green (Figure 0.2). Since then I have often delighted in observing the exquisite blue–violet shadows that can be conjured up simply by lighting a candle at sunset. But irrespective of how they arise, such intense coloured shadows have never failed to strike me as things of beauty and objects of wonder.

The large majority of those who first investigated coloured shadows of this sort were clearly impressed by the beauty of their colours. Indeed, the degree of consensus on this matter is nothing short of remarkable. In the earliest scientific paper to mention them, 'Dissertation sur les couleurs accidentelles' of 1743, Georges-Louis Leclerc, Comte de Buffon said that those he had observed in the open air at dawn and dusk were of a 'blue as vivid as the most beautiful azure'.[1] And in a later version of his essay, Buffon cited a remark about 'shadows of a very beautiful blue' from a letter which the Abbé Claude-François-Xavier Millot had sent him in 1754 or 1755.[2] The colour theorist, printmaker, and entrepreneur Jacques-Fabien Gautier, known as Gautier d'Agoty, declared in his *Chroa-Génésie* of 1750 that the colour of a shadow at the base of a candle towards dusk was an 'azure blue of the utmost beauty'.[3] In the first essay devoted exclusively to coloured shadows, 'Sur les ombres colorées' of 1767, Nicolas de Beguelin described their colour as a 'beautiful blue'.[4] Similarly, in the first book about coloured shadows, *Observations sur les ombres colorées* of 1782, a

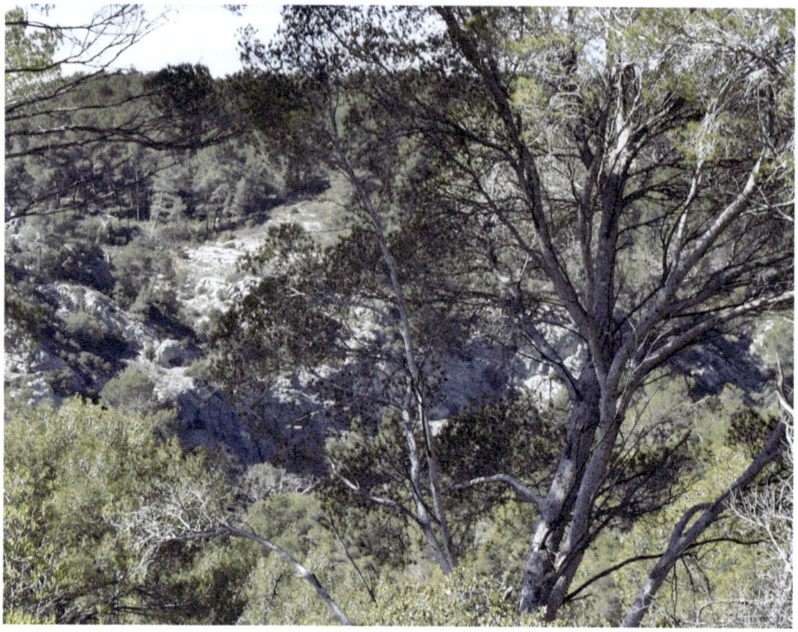

Figure 0.1 Blue shadows in the foothills of the Mont Sainte-Victoire, near Le Tholonet, Provence.
Source: Author.

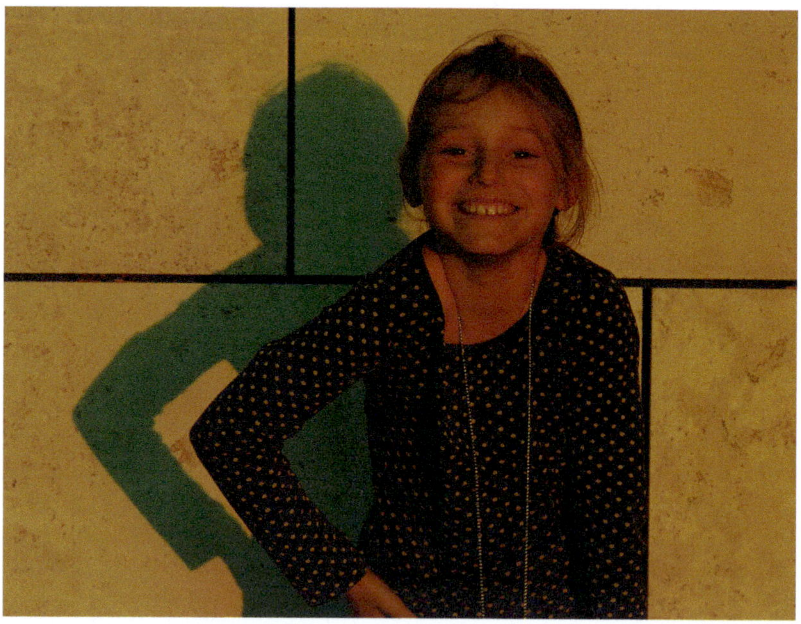

Figure 0.2 Miranda at the Getty, Los Angeles.
Source: Author.

certain H.F.T. repeatedly referred to their 'beautiful blue colour' and also their 'beautiful green' and even 'beautiful yellow' colours.[5] In the same vein, the mathematician Gaspard Monge mentioned shadows of a 'very beautiful blue', a 'very beautiful green', and a 'very beautiful red' in his groundbreaking paper, 'Mémoire sur quelques phénomènes de la vision' of 1789.[6] An entry in the diary Goethe kept while visiting the Harz mountains in December 1777 described green shadows 'of a beauty comparable to the green of an emerald'.[7] And in his essay, 'Von den farbigen Schatten', written in 1792–93, Goethe extolled their 'beautiful blue' colour.[8] Most eloquently, perhaps, Sir Benjamin Thompson, Count Rumford declared the colour of a shadow to be the 'most beautiful blue that it is possible to imagine' in his seminal paper, 'An Account of Some Experiments upon Coloured Shadows' of 1794.[9]

Artists were more reticent when it came to putting the beauty of shadow colour into words. Philipp Otto Runge nevertheless described the 'beautiful green' and 'beautiful bluish violet' colours of shadows in his essay, 'Gespräche über Analogie der Farben und Töne' of around 1810.[10] Later, the artist Frank Howard mentioned their 'beautiful blue tint' in his popular manual, *Colour as a Means of Art* of 1838.[11] And in a book of recollections, *Mnemosyne: Blätter aus Gedenk- und Tagebüchern* of 1848, Carl Gustav Carus referred to the 'most glorious blue shadows'.[12] Similarly, Ford Madox Brown described 'lovely violet shadows' in his diary for July 1855.[13] And as recently as 2007, the artist Spencer Finch effectively reiterated Goethe's opinion when he painted the photograph, *The Most Beautiful Blue (Goethe's Theory)*. It is nonetheless safe to assume that the many artists who, from the late eighteenth century onwards, struggled hard to capture coloured shadows did so because they were convinced of their beauty.

To suggest, as I do, that there is something marvelous about coloured shadows, is to imply that their appearance defies easy categorisation and resists assimilation to everyday experience. Fewer investigators remarked on this aspect of their phenomenology, but among those who did, Goethe described coloured shadows as 'beautiful and strange phenomena' in 'Von den farbigen Schatten'.[14] Later, Carl Gustav Carus described the 'bluish' and 'purplish' tones of snow in 'shade' as 'wondrous' in a letter of around 1820.[15] And Ford Madox Brown included 'violet shadows' among the 'wonderful effects' he saw one July evening in 1855.[16]

In *Shadows and Enlightenment*, Michael Baxandall summed up historical beliefs about the nature and causes of coloured shadows and their deficiencies. He argued, more particularly, that partisan commentators from the mid-eighteenth century onwards characterised the blue shadow either as a 'physical object' coloured by the light of the sky or as a 'product of the perceiving subject' generated by contrast, but never as both.[17] There are, in fact, numerous instances of a scientist or artist explaining a shadow's colour simultaneously as an effect of skylight and of contrast, which complicates his picture of a two-horse race between the two theories. Baxandall's other argument that such theories misrepresented coloured shadows because these are actually generated by constancy mechanisms in the brain is nevertheless correct.[18] The fact, however, that Gaspard Monge published a perfectly serviceable theory of coloured shadows based on constancy in a paper as early as 1789, which was ignored or misunderstood by most subsequent investigators, significantly complicates the historical situation.[19]

I also take issue with Baxandall's claim that coloured shadows 'are not very important to our understanding of the world'.[20] Or at least I will be concerned to demonstrate that tracing the evolving understanding of coloured shadows sheds light on cognition, since it illustrates how beliefs about their causes shaped how scientists and artists understood them. Arguably, these beliefs also affected how observers saw coloured shadows, so I will investigate the extent to which seeing them involved concepts or amounted to what the philosopher Wittgenstein called 'seeing as'.[21] And since the concepts at issue concern processes in particular, I will ask whether the seeing they engendered was tantamount to what Norwood Hanson dubbed 'seeing that'.[22]

In this context, it is worth mentioning that recent empirical studies demonstrate decisively that theories about their causes have a discernible effect on the understanding and perception of coloured shadows. One found, for example, that contemporary museum visitors employed a 'pre-existing model' or 'theory', normally of 'reflection', when asked to explain a display of coloured shadows or to predict the outcomes produced by varying the lighting producing them.[23] Similarly, another showed that subjects' knowledge of 'geometrical optics' contributed to their comprehension of coloured shadows and – conversely – that their ideas about paint mixture could inhibit it.[24]

If, historically, theories had a similar effect and sometimes misshaped the understanding and perception of colour shadows, they nonetheless facilitated these processes as well. At all events, it was only when Leonardo's ideas about coloured shadows came to the notice of French scientists in the mid- to late-eighteenth century that they began to appreciate that they were in fact vividly 'coloured' and not 'black' as the result of 'privation of light'.[25] Scientists began, in other words, to conceive of coloured shadows as positively illuminated areas of the kind Roy Sorensen now calls 'filtows'.[26]

One of the first to do so was Abbé Jean-Antoine Nollet, who cited Chapter 328 of Leonardo's *Trattato* on coloured shadows in its entirety in the fifth volume of his *Leçons de physique expérimentale* published in 1755.[27] Nollet also emphasised the importance of Leonardo's theory by stating that, even if it was one of those 'discoveries' in 'physics' which was 'forgotten and even lost' and only 'rediscovered after several centuries', it still invalidated Buffon's claim that 'no astronomer or physicist, nor anyone had spoken of the phenomenon' of coloured shadows before he did.[28] And to drive the point home, Nollet appended a footnote to his text which stated that the (second) 1716 edition of the *Traité de la Peinture* was 'very instructive, not only for painters but for physicists'.[29] Similarly, Pierre Bouguer noted pointedly, in his *Traité d'optique* of 1760, that the coloured shadows described by Buffon were a 'most singular phenomenon' to which 'painters had not failed to be closely attentive'.[30]

Not all early observers' perceptions were as 'theory-laden' as the foregoing account suggests, however.[31] From the historical point of view, that is, several scientists acknowledged the inability of the theories available to them to account for all the behaviours coloured shadows exhibited. And some, like Monge, even developed embryonic theories of constancy which could explain them more satisfactorily. The epistemic situation was further complicated by the fact that painters were not perceptually hamstrung by what they could glean about coloured shadows from science. Indeed, some – like Pierre-Henri de Valenciennes – were quite sceptical towards this form of explanation despite their considerable erudition, while others enhanced their perception of coloured shadows by inventing novel strategies and clever devices for viewing them.

Most of what follows will detail the evolution of the ideas and practices concerned. But I will end this book by returning to the issue that prompted it in the first place, namely what it is that makes coloured shadows so extraordinarily beautiful. One reason for doing so is that this topic has received no serious attention from philosophers, despite the weight of opinion on the matter. Obviously enough, this task will involve addressing the fact that coloured shadows are pleasing to look at, but it will also mean examining whether they are wonderful because – among other things – they can suspend the physical attachment to the material world we normally experience when seeing.

Inasmuch as this book is solely concerned with coloured shadows produced by two light sources, it does not dwell on the closely related shadows coloured by reflections from neighbouring objects, of the kind Vermeer seems to have observed. Neither, I must emphasise, is every shadow an artist depicts with colour intended to represent a coloured shadow. This is because a painter can use 'blue shadows' to indicate 'veins' under the 'eyes', as John White suggested in a manual for artisans of 1710.[32] Or the artist may employ 'blue' to depict the 'Turnings' or fleeing edges of objects, as Jakob Le Blon recommended in his *Coloritto* of *c.* 1725.[33] And sometimes, a painter will use blue to depict a colourless shadow blue because this creates a softer contrast than black.[34] I have therefore ignored paintings when there is no evidence that the shadows depicted in them were perceived as coloured. These include the Hellenistic wall paintings at Kazanlak in which one commentator discerned 'blue shadowing' akin to that in 'French Impressionism'.[35] By the same token, I have excluded works by Giorgione and Titian, which contain blue shadows according to Charles Eastlake, and the illuminations of Barthélemy d'Eyck, in which Philippe Lanthony has discerned 'tinted shadows'.[36]

The most recent artists I have included in this study are the Impressionists (including Cézanne), whose work I follow up to around 1900. I have excluded the so-called Post-Impressionists, Seurat, Gauguin, and Van Gogh, for all that coloured shadows feature prominently in their paintings, because they were less concerned with perceptual veracity than idealisation and exaggeration. (This is clearly attested by the advice Paul Gauguin gave to Paul Sérusier in 1888 to paint a 'bluish' shadow 'as blue as possible', using 'pure ultramarine'.)[37] By 1900, moreover, coloured shadows became a commonplace and even something of a cliché in painting.

As regards theories of coloured shadows, I have tried to touch on the most familiar and historically important texts produced by scientists and artists alike. But I have also devoted space to texts which have not received the attention they deserve, and, to some extent, to those as well which are almost unknown. By looking at the works of Karl Scherffer and Pietro Petrini, therefore, alongside that of Eugène Chevreul and Hermann von Helmholtz, I hope to offer a more historically replete account than available hitherto and to draw attention to some of the idiosyncratic and sometimes brilliant ideas advanced by the less well-known figures. The same goes for the artists discussed here, who include August Kopisch and Harry Willson as well as Ford Madox Brown and Claude Monet.

Notes

1. Leclerc 1746, 158. Cited in Baxandall 1995, 112.
2. Leclerc 1774, 337–338.
3. Gautier 1750–51, 2: 70.

4. Beguelin 1769, 33, 34, 37, and 40.
5. H.F.T. 1782, 16, 159, 161, and 217; 147 and 152; and 106. H.F.T is misidentified as Jean Henri Hassenfratz in Matthaei 1971, 49. See Roque 1994, 412.
6. Monge 1789, 135–136 and 146. The manuscript of this essay is preserved in the Fonds Gaspard Monge at the Bibliothèque de l'École Polytechnique, Cote: IX GM 27 Bib.: BCX CRH.
7. Goethe 1840, 35.
8. Goethe 1897, 101 and 103.
9. Thompson 1794, 108 and 116.
10. Runge 1840, 1: 169.
11. Howard 1838, 75.
12. Carus 1848, 302.
13. Brown 1981, 145 (entry for 21 July). Cited in Staley 1973, 39.
14. Goethe 1897, 101.
15. Carus 2002, 85.
16. Brown 1981, 145.
17. Baxandall 1995, 117.
18. Ibid., 116–117.
19. See Monge 1789.
20. Baxandall 1995, 117. Among the pioneering art-historical studies of the topic are Moran 1910; Reutersvärd 1950; and Staley 1973, 24–29 and 63. More recent studies include Charnay and Sieffert 2001; Lanthony 2006, 23–31; Schaeffer et al. 2008, 31–33; Fiorani 2008; Fiorani 2009, 125–135; Roque 2009, 41–48, 55–65, 71–84, and 89–108; Schöller 2017; and Smith 2018.
21. On 'seeing as' in this sense, see Wittgenstein 1995, 193e–214e; Hanson 1965, 20–21 and 24; and Fodor and Pylyshyn 1981, 188–193. For different interpretations of the phrase, see McGinn 1977, 190–195; and Good 2006, 7–33.
22. On 'seeing that', see Hanson 1965, 20–22 and 24–25.
23. See Allen 1997, esp. 716 and 732.
24. See Olivieri et al. 1988.
25. Leclerc 1746, 157. Cited in Baxandall 1995, 112.
26. Sorensen 2008, 155 and 168–180.
27. Nollet 1743–64, 5: 512–514.
28. Ibid., 5: 512; citing Leclerc 1746, 157.
29. Nollet 1743–64, 5: 512.
30. Bouguer 1760, 367. Cited in Baxandall 1995, 113.
31. See Hanson 1965, 19.
32. White 1710, 40.
33. Le Blon 1725, 8.
34. There is a blue shadow of this kind in the forehead in Raphael's *Angel Bust* of 1500–01 in the Pinacoteca Tosio Martinengo, Brescia. On the effect of the 'attribute' of colour on a mark's representational effect, see Willats and Durand 2005, 3, 6–7, and 17–18.
35. Benson 2000, 86 (and 98, 112, and 114). For the argument that the blue pigment used for shading at Kazanlak was 'not a colour but a darkener', see Bruno 1977, 79–87, esp. 83.
36. Goethe 1840, 370 (Eastlake's Note E); and Lanthony 2006, 120.
37. Denis 1912, 162; and Sérusier 1942, 42–43.

1 The Science

Why Constancy Matters

To appreciate why it is mistaken to regard coloured shadows as the products of reflection or contrast, it is necessary to recognise how they – like object colours – are generated by constancy mechanisms operating at different retinal and cortical levels in the brain. (That said, the issues here are quite technical, so the reader who is not scientifically minded may prefer to skip this brief chapter.)

Constancy is so-named because it ensures that the colours of surfaces remain broadly the same in absolute terms across small changes in the colour or spectral composition of the illumination, and remain the same relative to one another across more significant changes.[1] Thus, a white wall looks white, or comparatively so, under both artificial and natural lights, despite the fact that the former is warm while the latter is much cooler. This way of stating the problem is intuitive or phenomenological. Objectively, however, things are the other way round. That is, in terms of colour temperature or the colour of a black body plotted against the line through colour space (known as the Planckian locus) that maps the notional chromatic changes it undergoes when heated, candlelight is normally around 1800° Kelvin, whereas sunlight and overcast daylight are around 6,500° K and the blue light of the sky can exceed 20,000° K.[2]

Although constancy works well when the illumination is relatively homogeneous and unsaturated and the eye has a chance to adapt, it breaks down when a scene is lit by a strongly coloured single light source. Or put simply, there is a point when things start to look red under red light, even white surfaces like the marble walls in the photograph of Miranda. Constancy also miscarries when a scene is illuminated by two light sources of different colour temperatures, when it will generate coloured shadows in those areas where only one of them reaches. As Rumford put it in 1794, shadow colour is thus a function of a 'difference' in the 'whiteness' of the illuminations in play in a scene.[3]

Comparison and Adaptation

In his explanation of the role played by constancy in producing coloured shadows, Baxandall refers briefly to ideas about 'lightness ratios', which Edwin Land advanced in a famous, and highly controversial, article of 1977, 'The Retinex Theory of Color Vision'.[4] These ratios refer to so-called triplets or the relative amount of long-, medium-, and short-wavelength lights reflected by the different areas of a visual field (at their edges), compared with each other.[5] Using assemblages known as 'Mondrians' made of pieces of coloured paper, Land found that these triplets accurately mapped the coordinates of their various colours in the three-dimensional colour space he devised. They could also predict the colour of shadows. Land demonstrated this last finding in an experiment that involved superimposing the beams of red and white lights emitted by two projectors.[6] By then placing a hand in the red beam and measuring the 'reflectance' of the resulting shadow, he obtained a triplet which indicated that its colour should be 'blue–green', which in fact it was.

John Mollon advanced a comparable argument in an essay of 2006, where he pointed out that Monge had effectively anticipated Land's ideas in the paper of 1789 mentioned earlier.[7] As Mollon explains, Monge illustrated his argument by drawing attention to the fact what while the shadow cast by an object on a piece of white paper looks blue when the scene is illuminated simultaneously by 'blue' daylight and a warm candlelight, the entire sheet of paper – including the area previously occupied by the shadow – looks 'white' when it is lit by the same daylight on its own.[8] Monge did not explicitly explain this phenomenon in terms of constancy, but Mollon did by extrapolating from his predecessor's more general theory, which implies that we estimate the colours of the different surfaces in a scene according to the 'ratios' of the different kinds of lights they reflect.[9] By this account, the shadow looks blue under a mixed illumination where the candlelight is blocked because it reflects the same flux of wavelengths as a blue object located elsewhere in the scene would if illuminated by both light sources. It is blue, in other words, not because the quantity of short-wavelength light it reflects is large in absolute terms, but because this kind of light constitutes a large proportion of the flux it reflects compared with the fluxes reflected by the objects around it.

Several recent commentators have pointed out that mechanisms exist in the higher reaches of the visual cortex which can implement this kind of analysis of the radiant energy reflected by a scene. All such mechanisms are dependent on information provided by the primary visual cortex – what scientists call colour 'contrast', which in this case simply means variation or difference.[10] But the more particular differences between the colours of shadows and those of the surfaces around them are computed by high-level, long-range, or 'global' mechanisms in the cortex, which are responsible for identifying the chromatic contrast between one area of a scene and those further afield.[11]

Our perception of coloured shadows is also supported by the lower-level process of adaptation, which is implemented most powerfully within cone cells in the retina. Here, it serves to ensure that the level of excitation attained by neighbouring cones of the same spectral sensitivity is reduced in inverse proportion to the level of excitation (or according to the Von Kries coefficient) when they are all stimulated by light of the same wavelength.[12] With adaptation, therefore, a white surface under slightly coloured illumination will produce much the same level of excitation in these cells as a surface under a neutral or white illumination would. This helps explain Monge's conundrum that white paper lit by faintly blue daylight looks white. The same principle also implies that a white surface lying under pure white light will appear slightly cooler in colour to the cones that 'see' it, when its surroundings are illuminated by a faintly coloured illumination adjusted to white. Adaptation can therefore explain why a shadow illuminated only by neutral daylight looks blue nonetheless, when it is enclosed by a physically and apparently white field illuminated by a mixture of daylight and candlelight.

Adaptation also explains why white surfaces are of especial importance to constancy, and shadow colour, with the idea that they act as a benchmark against which we can compute the colours around them. (It is no coincidence, therefore, that sheets of white paper and snow occur over and over again in accounts of coloured shadows.) In his 1977 article, for example, Land demonstrated that white serves as a control when we compare reflectances (although the system he devised here did not actually require it).[13] This makes sense because comparing the manifold reflectances of coloured objects with that of a standard white is a much simpler computation than comparing the reflectances of every coloured object in a scene with one another.

The visual system can also identify white easily since it is always the most luminous matt colour in a scene.[14] And perhaps most importantly, a white surface is chromatically invariant across its whole surface under a homogeneous illumination, even in shadow (whereas other colours change hue slightly when lightened or darkened).[15] When, therefore, we use the white illuminated by a combined illumination to establish constant colours, as if the illumination were homogenous, this generates colours in the shadows where the illumination is different. And the same is true for those digital cameras which use a comparable constancy mechanism for achieving white balance, which means that coloured shadows can be successfully captured in photographs.[16]

Frederic Leighton's *View in Capri* of *c.* 1859 (Figure 1.1) gives some sense of the role played by white in generating coloured shadows. Not only does it depict blue shadows on the walls illuminated solely by the diffuse light of the sky and warm shadows on the walls lit only by the direct light of the sun, it also represents the

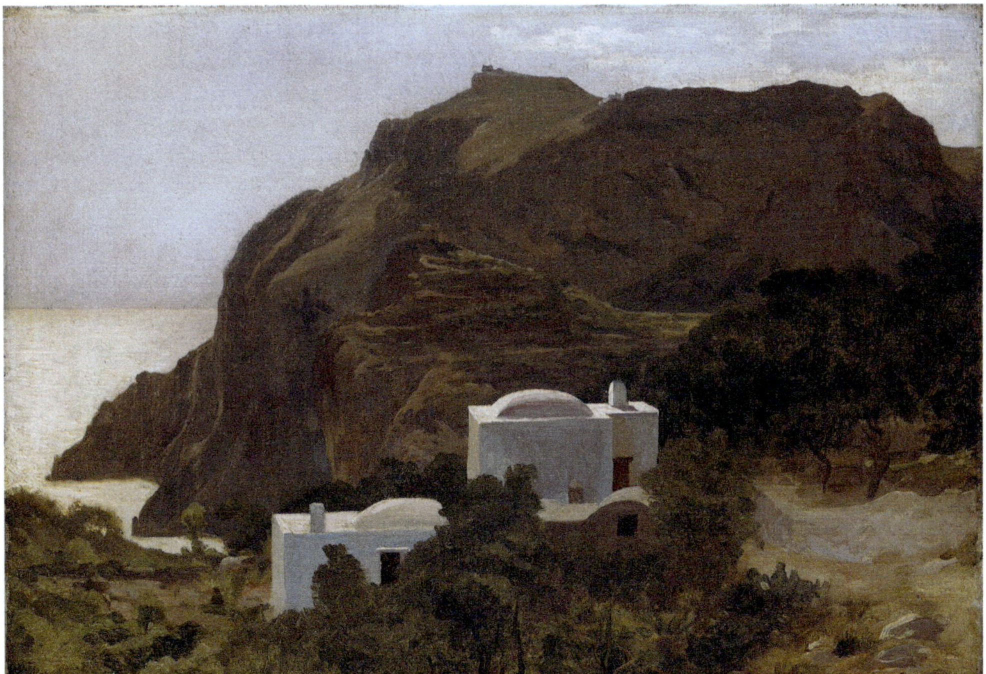

Figure 1.1 Frederic Leighton, *View in Capri*, *c.* 1859. Oil on canvas, 21.2 × 29.3 cm. Gere Collection, on long-term loan to the National Gallery, London, inv. no. L847.

Source: © Private collection 2000.

horizontal surfaces where both light sources reach as white – or very nearly, since constancy is never perfect and particularly (as already mentioned) under such circumstances when the light is strongly coloured.[17]

Some Caveats

None of this is to suggest that the saturation of the coloured light falling on a shadow makes no contribution to its appearance, but merely that any effect it does have is ancillary to that of constancy. So, although the sky does not colour a shadow in the same way as a buttercup colours the chin it is held underneath, a sky which is sufficiently blue in the right place can generate shadows which appear coloured even in bright sunshine by comparison with its putatively white light.

Simultaneous contrast, which occurs when two differently coloured but contiguous areas are viewed at the same time, can also intensify shadow colour. However, this phenomenon is produced by 'short-range' mechanisms, which use lateral interactions between retinal neurons capable of sampling chromatic differences between adjacent edges, to cause these to induce their 'opponent' complementaries in one another.[18] In the opponent colour system, first proposed by Ewald Hering in 1878 and subsequently developed by Leo Hurvich and Dorothea Jameson in the 1950s, red and green, and yellow and blue, are the mutually exclusive complementary pairs.[19] Simultaneous contrast therefore generates colours slightly different from those produced by constancy, which normally adjusts warm or cool variations in the illumination to white along a line through colour space running close to the daylight (or Planckian) locus, although it may also involve adjusting the totality of colours in the scene to grey.[20] The mechanisms responsible for producing simultaneous contrast are assisted by others which produce perceptual 'fill', which means that contrast can induce a faintly blue or blue–green colour in a white area surrounded by one of a warmer colour across its entire extent.[21]

Although numerous investigators invoked simultaneous contrast in their explanations of coloured shadows, the fact remains that it is simply too weak to produce colours of the same intensity and saturation as those generated by constancy.[22] Similarly, reflection can only colour a shadow to a significant extent when it is so abundant that it overwhelms the dominant illumination, which rarely happens. In the last analysis, then, only constancy mechanisms can account for what is most essential to the phenomenology of shadow colour: that it is not dull or wishy-washy but luminous and saturated.

As a function of constancy, the colour of a blue shadow is neither an 'objective' physical imprint, nor a reflection of the colour of the sky, nor a 'subjective' contrast effect either.[23] Indeed, if the chief purpose of constancy is to regulate the subject's perception of, and engagement with, the coloured surfaces of the objective world, none of the effects it produces – and coloured shadows among them – can be assigned exclusively to either realm. I will return to this issue in my conclusion, where I consider what makes these effects 'beautiful'.

Notes

1. For a review of the neuroscientific scientific literature, see Foster 2011. For a philosophical overview of constancy, see Matthen 2010.
2. See Davis 1931; and Borbély et al. 2001. See also Hire 1694, 239, on how to make the colours of different illuminations apparent by viewing them simultaneously.

3. Thompson 1794, 110. See also ibid., 111, which describes how (by corollary) the light from two lamps filtered through the 'same yellow glass' produced shadows 'without the least tinge of colour'.
4. Land 1977.
5. Land and McCann 1971, 4 and Land 1977, 12, 15, and 17.
6. Ibid., 16.
7. Mollon 2003, 21 and Mollon 2006, 299.
8. Ibid., 299–300; citing Monge 135 and 145–147.
9. Mollon 2006, 301. See also Foster and Nascimento 1994.
10. See Brou et al. 1986.
11. See Hurlbert 2003.
12. See Foster 2011, 679 and 692 and Kallmann et al. 2014, 193.
13. Land 1977, 12–13. See also Land 1983, 5167 and Hardin 1988, 191.
14. See Land and McCann 1971, 3 and Foster 2011, 676–677.
15. See Land 1977, 12; and Mollon 2006, 306.
16. See Sik-Lanyi et al. 2019, 888–871, for a comparison between human constancy and the 'white balance' mechanism of a digital camera.
17. See Evans 1948, 130–131, for the claim that 'Adaptation is almost never complete' since 'There is always a residue which permits the observer the see the direction of the departure from white of any given stimulus.' See also Pearce et al. 2014, 10 and Webster 2015, 208.
18. See Hurlbert and Wolf 2004 and Kallmann et al. 2014, 192–194. On the physiological basis of simultaneous contrast, see De Valois and De Valois 1997, 132–136.
19. See Hering 1964, 49–50; Hurvich and Jameson 1957. See also Hardin 1988, xi and 28–36.
20. On the first kind of adjustment, see Mollon 2006, 307; Webster and Leonard 2008, 2818; and Pearce et al. 2014, 1 and 6. On the 'grey world' assumption, see Hurlbert 1998, 298.
21. See Kallmann et al. 2014, 192.
22. See ibid., 186. For other weakness in the argument that coloured shadows are generated by simultaneous contrast, see Webster et al. 1988 and Spillmann and Werner 1996. For the claim that coloured shadows are a function of simultaneous contrast, see Delabarre 1889 and Valberg 2005, 262. See also Churma 1994, 4721, which describes 'simultaneous contrast' as the phenomenon (which is in fact adaptation) that occurs when 'nerve connections beneath the surface of the retina' to the 'adapted receptors around them', 'desensitised' to the 'yellow light' falling on a shadow's surround, make 'any bluish light in the shadow all the more visible'.
23. For an early discussion of the 'objective' and 'subjective' coloured shadows (supposedly) produced by blue light and contrast, respectively, see Müller 1840, 2: 418–421. For philosophical discussions of subjectivity and objectivity in the context of colour constancy, see Hardin 1988, 82–91 and Cohen 2009, 53–57.

2 Light and Dark

Leonardo on Causality

The earliest accounts of coloured shadows are to be found in the writings of Leonardo da Vinci, who investigated them both as an artist and as what we would now call a scientist (the division between the two roles only being institutionalised in the seventeenth century). Indeed, the first edition of Leonardo's notebooks – drawn from copies Franceso Melzi made of them around 1540–50 and published in 1651 as the *Trattato della pittura* – contains no fewer than five chapters devoted to coloured shadows, as they appear at sunrise and sunset, and in daylight, on white surfaces.[1]

Leonardo's language is sometimes a little obscure.[2] It is nonetheless clear that the ideas he developed about the causes of coloured shadows shaped how he saw them, as areas coloured by the light of the sky. Coloured shadows are not so simple, of course. So, at the same time as concepts gave a definition to Leonardo's investigation, they produced a degree of 'aspect-blindness' in him towards their actual causes.

Just how ideas inflected Leonardo's way of seeing can be appreciated by considering Chapter 27 of the *Trattato*. This states, with respect to the 'shadows' of 'figures' in the 'countryside' at dusk, that:

> These shadows will be poor companions for the illuminated areas, because the blue of the air will illuminate that [shaded] side, tinting the part that it strikes with blue. This is very obvious in the case of white things [when the] blush of clouds, together with the flush of the sun, gives a red tint to whatever takes light from them, and the side of bodies not facing this redness remains the colour of air.[3]

This much implies that shadows at sunset were coloured by the skylight reaching them in the same way as illuminated surfaces were coloured red by the setting sun.[4] Leonardo subsequently refined this argument in Chapter 328, where he explained 'Why shadows cast by bodies onto a white wall at the onset of evening are blue', as follows:

> The shadows of bodies produced when the sun is reddish near the horizon always appear blue. This occurs according [to the proposition that] the surface of every opaque body takes on the colour of the object facing it. Therefore, given that the whiteness of the wall is entirely devoid of any colour, it becomes tinged with the colour of the objects facing it, which in this case are the sun and the sky. Because the sun turns reddish towards evening and the sky remains blue, wherever shadow

> does not face the sun . . . it will face the sky. Therefore . . . this derived shadow will be struck by the blue colour [reflecting] off the white wall, and the field around that shadow facing the reddish sun will take on the colour red.[5]

What this passage adds to the previous one is the idea that the sky transfers its own colour to a shadow with the light it emits. It thus explains how shadows acquire their colour in terms of the more familiar process whereby an object reflects its colour into the shadow of another object nearby, which Leonardo described elsewhere.[6]

Unfortunately, Leonardo's ideas on this matter were somewhat obscured by the engraving (Figure 2.1) employed in the first edition of the *Trattato* to illustrate Chapter 328. The problem with this is that it misconstrued the drawing upon which it was based, Melzi's copy of Leonardo's original, which is quite difficult to interpret as an illustration of the shadow cast by an object suspended from a rod which has been cantilevered out from the top of a wall.[7] More particularly, the engraving transformed the arc in Melzi's drawing, denoting the vault of the sky, into a proscenium arch with the result that much of the information in the original about the geometry of light was lost.[8] The engraving also introduced a welter of irrelevant detail, making it too specific to serve the purpose of illustrating a theory efficiently.[9]

Obviously enough, a monochrome image is not well-suited to depicting the production of shadow colour.[10] By dint of being coloured, however, a painting can exemplify shadow colour in a more replete fashion and thereby make the artist's ideas about its causes explicit. Francesca Fiorani has argued that Leonardo did precisely this in introducing blue shadows of the kind he described in Chapter 328 into his *Annunciation* of *c.* 1472–75 (Figure 2.2).[11] These, she states, are tinged with blue representing the 'indirect, blue light of the sky vault', while the illuminated faces of objects are tinged with yellow representing 'the direct, red light of the sun'.[12] There is little doubt that – just as Fiorani observes – the thin paint in the rusticated masonry on the right of the Virgin allows the warm ground below to show through, giving the stonework here the warmish colour of the setting sun, while the blocks lying in shadow to their left look 'bluish' like the sky.[13] Nor can it be denied that the wall enclosing the garden in the

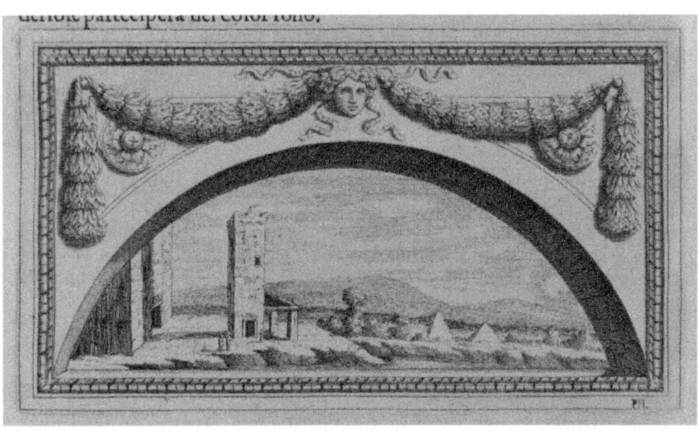

Figure 2.1 Blue shadows. Illustration to Chapter 328 of Leonardo da Vinci, *Trattato della pittura* (Paris: Jacques Langlois, 1651b). Engraving by René Lochon.

Source: Author.

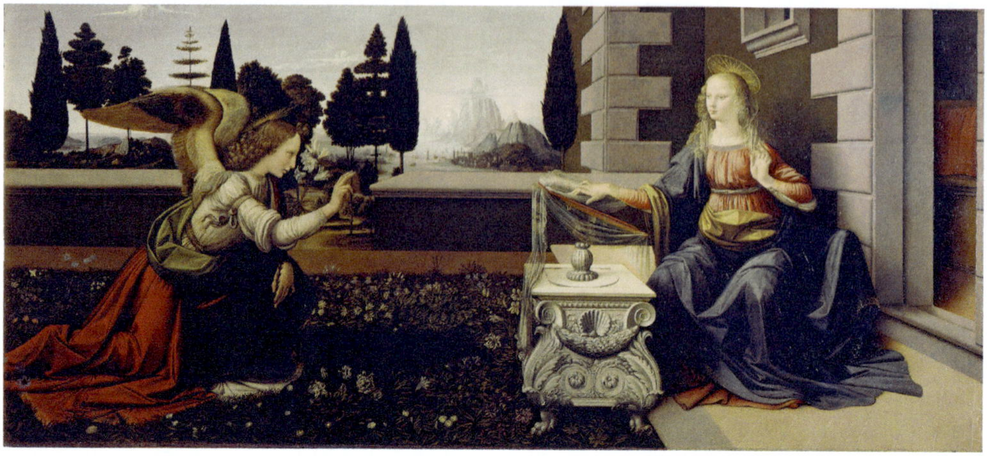

Figure 2.2 Leonardo da Vinci, *Annunciation*, *c.* 1472–75. Oil on panel, 90 × 222 cm. Florence, Uffizi, inv. no. 1890, 1618.

Source: Uffizi.

background is, as Fiorani says, 'yellowish' towards the left where it is 'closer to sun' and 'more bluish' towards the palace', where it is 'more exposed to the sky vault'.[14]

Elsewhere, however, things are less clear-cut. There are, as Fiorani claims, a 'yellowish' cast to the top surface of the base of the lectern at which Mary is reading where it is open to the warm light of the 'late afternoon sun' and, arguably, a 'bluish' tinge to its front face where it is touched by the 'blue light of the sky'.[15] Fiorani points out, however, that the chromatic difference between the two surfaces is enhanced by the present varnish.[16] And while the quite thick white paint used to depict the raised border of the same surface does tend towards yellow particularly towards its top and right edges, it may do so as a result of ageing in the oil binding the paint. There is also a discernible warmth in the area enclosed within this border, and in the circular section especially, but this could be an unforeseen result of how the paint here allows the pink underpainting to show through. The front face of the base does look cool from up close, but largely because a measure of blending between the layers of black and white paint Leonardo used to create relief in the ornamentation results in a faint, and presumably accidental, bluing in places. It also appears quite grey from a distance, when gauged against a standard Kodak grey card.[17]

It seems likely, then, that the convention that shadows were colourless prevented Leonardo from making them as richly saturated as his written observations indicate that he believed them to be. Another explanation of their subdued colour is that Leonardo's commitment to what we nowadays call a scientific method diminished his sensitivity to their phenomenology. There is a case to be made, in other words, that Leonardo was less attentive to shadow colour *per se* than its causes because he preferred a form of painting which favoured explanation over description. This argument gains credence from the fact that he admonished the painter to make the 'cause' of 'shadows and lights' explicit in Chapter 27 of the *Trattato*, where he also told him that 'your work will be empty and false if you do not make them partake of the causes mentioned above'.[18] Similarly, Leonardo advised the 'painter' in Chapter 6 of the *Trattato* to not only to observe the 'objects appearing in front of him', but also 'consider how they fit the rules concerning place, circumstances, lights, and shadows'.[19]

In Chapter 7, Leonardo made the recommendation to the painter to 'First, study science and then pursue the practice born from that science', adding that a 'painter must study with discipline, and must not leave anything uncommitted to memory'.[20] Taking this together with his other injunctions, it is clear that Leonardo wanted the painter to work from explanatory formulae which were derived from observation and then committed to memory. This approach also diminishes the observer's sensitivity to the phenomenology of shadow colour, first of all, because our memories of colours assimilate them to semantic categories, which – like words – cannot capture the 'fine-grained' quality of our visual experience.[21] Second, as Ewald Hering argued, memory reinforces our proclivity to disregard 'incidental colours' of the kind belonging to 'shadows' and favours our tendency to see the 'real colour' of an object in terms of the particular 'memory colour' we normally attribute to its surface.[22]

There is more to Leonardo's understanding of the causes of coloured shadows than the belief that they were coloured by the light of the sky. This is not to suggest that Baxandall was mistaken in arguing that Leonardo's theory is consistent with Pierre Bouguer's, much later, 'physicalist' account of their origins.[23] Rather, it is to suggest that this claim does not capture how Leonardo envisaged the more fundamental process responsible for colouring the sky in Aristotelian terms as the effect of the process, whereby the white of the earth's atmosphere becomes mixed with the black of space behind it.[24]

Leonardo stated this belief in Chapter 151 of the *Trattato*, where he argued that 'The blue of air arises from the thick body of illuminated air between the tenebrous [region] above and the earth'.[25] And in Chapter 134, he explained in more detail how, 'as the species' of 'total darkness penetrate through the air, the whiteness of the air will appear blue to us'.[26] These passages in fact rehearse an argument advanced in the *De Coloribus* written by a member of the early Peripatetic school, possibly Theophrastrus or Strato, around the middle of the fourth century BCE, which he may have read in manuscript around 1508 (before its publication in Latin in 1548).[27] More particularly, *De Coloribus* described how, 'when examined in depth', the air appears 'to be blue in colour' where the 'light fails' and it is 'penetrated by darkness'.[28] For Leonardo, then, the sky appears blue just as 'dark' paint is made to look 'blue' when a layer of transparent 'light' paint is superimposed over it, as he described in Chapter 113 of the *Trattato*.[29]

In other passages of the *Trattato* concerned with the production of blue, it is apparent that Leonardo did not regard darkness simply as the privation of light but as a positive principle on a par with it. This is clearly the case in the explanation Leonardo gave in Chapter 302 of how the mixture of light and dark generates the colour of the sky. This sets out how:

> [A]ir takes its light from the sun and total darkness from the absence of that sun. . . . [M]oisture mixes with it from the middle region downward, thickens it, and after it thickens, solar rays strike it and illuminate it. The air above the middle region remains tenebrous. And because light and total darkness compose the colour blue, this blue colours the air with greater or lesser darkness as that air mixes with greater or lesser humidity.[30]

It is fair to conclude, then, that Leonardo conceived of the ultimate causes underlying the production of coloured shadows in antiquated terms, even if he understood their immediate causes in terms which anticipated modern physics. Most importantly, perhaps, he did so without recognising any significant incommensurability between them.[31]

Later Aristotelian Explanations

There was little interest in coloured shadows among artists or scientific investigators during the next 150 years. At least, they were not discussed in any published text before the *Lectiones XVIII* published in 1669 by Isaac Barrow, Newton's predecessor as Lucasian Professor of mathematics at Cambridge. Here, Barrow took an Aristotelian stance on the nature of blue, arguing that it was produced by the mixture of 'alternate black and white particles'.[32] He thus declared that 'white objects' will 'occasionally show this colour when poorly lit up' or when light is compromised by dark.[33] He also argued that the 'clear sky' looks blue because this contains 'only a few corpuscles' that are capable of 'beating back the light to our eyes, while the rest of the light slips through'.[34]

In addition, Barrow stated that the 'shadow of any opaque object cast, by day, by a burning lamp and caught or terminated on a white paper' is blue by the same token or because it is a mixture of light and dark.[35] In order 'to be explicit' about what he meant, he supplied the first diagram to illustrate the joint operation of artificial and natural light in producing coloured shadows (Figure 2.3).[36] What this shows is that when the candle or lamp at C shines on the object extending from A to B, it casts a shadow on the white paper from X to Y illuminated by daylight. Implicitly, therefore,

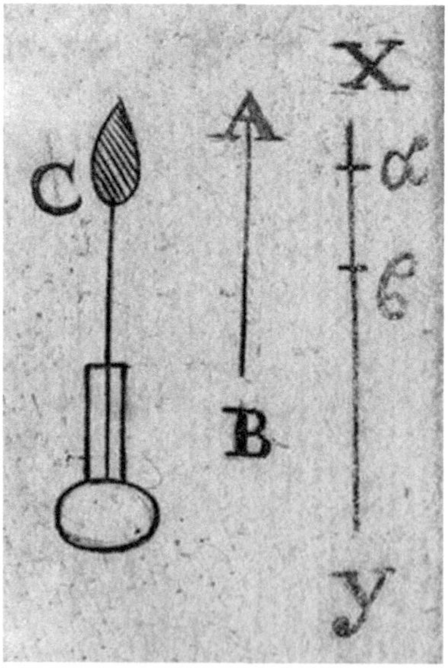

Figure 2.3 Coloured shadows. Fig. 128, Plate 10 from Isaac Barrow, *Lectiones XVIII* (London: William Godbid, 1669).

this shadow is blue between α and β, because its darkness is attenuated by weak daylight in this segment.[37]

Barrow's account has been largely ignored, unlike the slightly later account Otto von Guericke gave of the coloured shadows he observed in the same circumstances in his *Experimenta Nova* of 1672.[38] In this, and like his predecessor, Guericke explained the process of producing the blue colour of shadows by analogy with other processes involving the mixture of light and dark, contending that:

> Between black & white . . . is the colour blue. Just as a drop of milk and a drop of the ink will produce a blue colour where they mix, so in the morning, at dawn, the blue shadow on a flat sheet of white paper produced when a finger . . . is placed between the candle and the paper . . . does not appear black but a perfect blue colour.[39]

Aristotelian explanations of colour and light began to drop out of favour after the publication of Newton's *Opticks* in 1704, which seemed capable of explaining an object's colour as a direct function of the quality and quantity of 'homogeneal' rays it reflected.[40] Despite their explanatory power, however, Newton's ideas were fiercely contested by French scientists.[41] They were also refuted by the committed Aristotelian printmaker and colour theorist, Gautier d'Agoty, who subtitled his *Chroa-Génésie ou génération des couleurs*, first published in 1749, *contre le système de Newton*.[42]

In one passage of the enlarged, second edition of this work published in 1750, Gautier described the coloured shadow he saw by chance as he observed a candle at dusk, thus:

> One evening . . . I noticed that the bottom of the candle that illuminated the space I occupied was of an azure blue of the utmost beauty, and that this was because the window was open and it was very dark, so, as the flame was interposed between my eye and the window, I saw the shadow beneath this flame through it, which produced such a beautiful blue for my eyes because of its dimness in this section.[43]

This passage is typically opaque; but what Gautier meant was that the shadow appeared blue because it was an area of darkness seen through light. This claim can be more fully understood in terms of his more general theory of colour, which characterised darkness as an active principle (just as it was for Leonardo).[44] 'Shadow', more specifically, was the property of 'repose' among particles and not simply 'nothing' and, hence, could create colour by modifying the level of 'activity' in light.[45] Dark seen through light or 'light interposed to Shadow', therefore, made 'pale and dark blue'.[46]

For Gautier, the sky's 'celestial colour' was produced in this way 'during the day', when the 'black and shady air' of the distant sky is subject to the 'interposition of rays between us' and it.[47] The blue colour at the base of the candle was produced by the same process, namely the 'interposition of light over shadow'.[48] Gautier believed he confirmed the cause of this 'phenomenon' when he suppressed the colour of the shadow by blocking the darkness from the window with a 'sheet of white paper' placed 'beyond the flame, and directly opposite it'.[49] He also took it as confirmation of his theory that a 'light blue' and then 'dark blue' reappeared when he moved the paper away.[50]

Gautier did not suggest in this passage that blue skylight coloured the shadow. In his *Observations sur l'histoire naturelle, sur le physique, et sur la peinture* of 1752, however, he did argue that the shadows visible in the open air at dawn and dusk were coloured by 'reflections', which in eighteenth-century parlance often referred to light from the sky. This described how:

> Sunlight is directed at object from only one point. . . . But universal light . . . comes from the same side and several points too, as in a landscape before sunrise and after sunset, [when] its western or eastern part is the most highly illuminated, and the opposite side the most shadowy, but in such a way that the great quantity of reflections, different from the light of the sun, which reverberate in all directions depending on their incidence, surround the object and gently light its shadowed parts.[51]

Irrespective of the precise meaning of this passage, in Gautier's mezzotint, *Apollo or the Sunrise* of 1743 (Figure 2.4), it does look as though the sky supplies the distinctly blue tones visible in the shadows opposite the warm tones that the rising sun gives to the god's flesh.

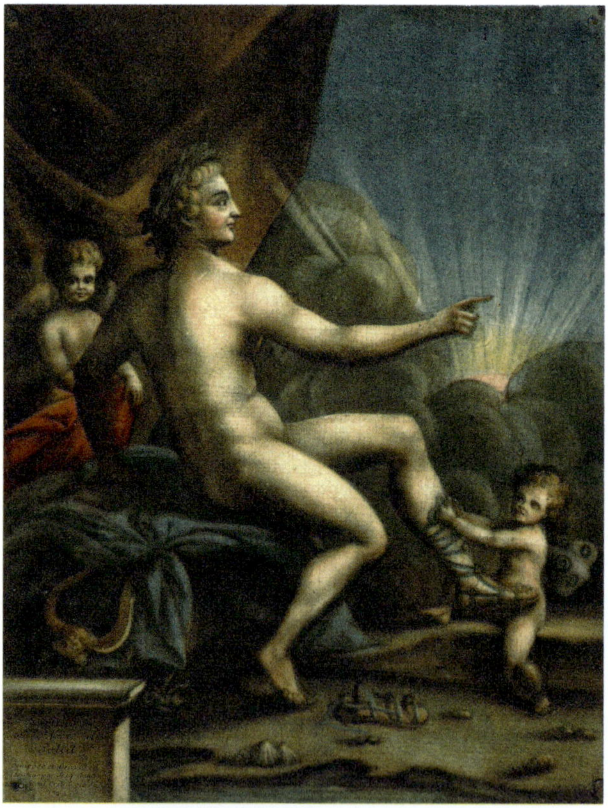

Figure 2.4 Jacques-Fabien Gautier d'Agoty, *Apollo or the Sunrise*, 1743. Mezzotint, 31.2 × 23.7 cm. London, British Museum, museum no. 1857, 0613.330.

Source: © Trustees of the British Museum.

Despite their anachronism, Gautier's ideas were rehearsed by H.F.T. in his *Observations sur les ombres colorées* of 1782.[52] The physicist did this because, after performing an exhaustive series of experiments into the shadows generated by combinations of direct, reflected, and filtered natural and artificial light upon neutral and coloured surfaces alike, he was forced to concede that the colour of shadow was not always a straightforward function of the light illuminating it.[53] For example, H.F.T. was at a loss to explain how a shadow could – he claimed – appear 'blue' on a white wall when this was a 'reflection' of the light of a 'candle' which emitted a 'perfect light composed of all the primitive colours'.[54] Nor could he make sense of how a uniformly white wall reflected a 'yellow' light from one part of its surface and 'blue or violet' light from another 'according' merely 'to the quantity, quality, and arrangement of the lights illuminating it'.[55] His puzzlement related, in particular, to experiments he had performed with two candles and no other light source, which – he maintained – produced 'yellow' shadows and 'violet–red' shadows (varying in colour between dark 'violet–purple', 'pale red mixed with violet', and a 'yellowish').[56] H.F.T. also claimed that he could alter and even switch the colours of the shadows produced by these candles simply by changing their position, thereby altering their 'intensity' relative to one another.[57]

H.F.T. asked whether Gautier offered a way out of this impasse with the idea that 'every colour' is produced by the 'mixture' of 'light and shadow'.[58] In expanding on this thought, he mentioned several aspects of his predecessor's general theory of light.[59] He discussed Gautier's argument that 'blue' was produced by 'interposition' or the 'transparency of light over shadow'.[60] And he examined the corollary of this claim that 'red, orange, or yellow' are 'produced by *the interposition of Shadow over light*'.[61] H.F.T. then concluded that, taken together, Gautier's different ideas about the 'concourse of light and dark' appeared at first sight to be capable of explaining the 'colours of Shadows'.[62]

Even so, H.F.T. admitted that Gautier's theory was not 'conclusive'.[63] He argued, more particularly, that it could not explain why the shadows cast by a chair opposite a cross-window switched colour when the light shifted, so that the set cast by the upper window was blue at 11.30 a.m. and yellow at 1 p.m., while the other set was first yellow and then blue.[64] This simply could not happen, H.F.T. reasoned, according to Gautier's theory that blue is produced by the superposition of light over dark and yellow is produced by the superposition of dark over light, since nothing had occurred, which could invert the relative ascendancy of the two light sources. Similarly, H.F.T. declared that Gautier's theory could not explain the results of related experiments which he had performed with firelight and candlelight.[65] So, although H.F.T. had previously asserted that the colours of shadows depended on the 'greater or lesser brightness generated by the light sources' producing them, he was forced to conclude that the 'nature' and 'properties of light' were yet to be fully 'understood' by physics.[66]

The scientist nevertheless returned to the idea that colour is generated by the mixture of light and dark later in his book, in an attempt to make sense of other shadow colours which the theory of reflection could not explain. He pointed out, for example, that different light sources could produce the same colour in the shadow they illuminated, provided the one casting it was altered.[67] Thus, the shadow lit by the atmosphere and cast by a candle was blue, but so too was the shadow lit by a candle and cast by firelight. Conversely, H.F.T. observed that the same light source could generate different colours in the shadow it illuminated, when the light casting it was changed.[68]

So, for example, the shadows illuminated by a candle varied between yellow, blue, and violet, as the light source casting them was changed from moonlight, to firelight, and to candlelight.

H.F.T. concluded from all this, quite correctly, that the light source casting a shadow must 'contribute' to its colour.[69] And he believed this insight could explain the anomalies he had mentioned earlier, since it was now apparent – he continued – that as the light source casting a shadow became stronger or weaker', it modified the colour of the shadow concerned.[70] H.F.T. consequently formulated the law that the colour of a shadow was a function of the relative 'brightness' of the source illuminating it compared with that of the source casting it.[71] More specifically, he argued that although the light casting a coloured shadow is always weaker than the light illuminating it, shadow colour varies according to quantitative difference between their relative strengths. Thus, a shadow is yellow when the difference is greatest, red when it is slightly smaller, violet when it is smaller still, and blue when it is at its smallest.[72]

What actually produces shadow colour is of course the difference in colour temperature between the light sources, and H.F.T. did accept that the 'colour' of a light source affected shadow colour.[73] But in order to preserve his Neo-Aristotelian theory, he argued that chromatic differences in light were themselves a function of differences in 'energy', white being the colour produced by the most powerfully combustible light sources and red the one produced by the weakest.[74] And although H.F.T. admitted that there were cases which this theory could not explain, he suggested that further research would illuminate them.[75] He therefore concluded that the colour of a shadow was nothing other than a function of the 'proportion of intensity' between the light sources casting and illuminating it.[76]

Goethe or Shadow Illuminated

Georg Lichtenberg made Goethe aware of H.F.T.'s 'important' volume in a letter of October 1793.[77] Lichtenberg sent this in response to receiving the manuscript of the great polymath's essay, 'Von den farbigen Schatten', originally intended for the second volume of his *Beiträge zur Optik*.[78] Goethe's essay described a number of experiments closely similar to H.F.T.'s, which called into question the notion that a coloured shadow is simply a 'reflection' of the light source illuminating it, not least because the two were often different in colour.[79] Goethe demonstrated, for example, that the shadow cast by a candle was vividly 'blue' even when it is illuminated by 'grey' daylight; and he cast doubt on the idea that the shadow illuminated by a candle is always yellow.[80] Perhaps, then, Lichtenberg mentioned H.F.T.'s *Observations* in an attempt to shield Goethe from accusations of plagiarism. This would also explain why he pointed out that the book had been summarised by Johann Gehler in his *Physikalisches Worterbuch* of 1790.[81]

Goethe nevertheless maintained that he had arrived at his ideas independently. In November 1793, for instance, he wrote to Friedrich Jacobi, asking him to:

> [E]nd my treatise on coloured shadows to the Princess Galizin, with the following note at the end: 'In a French work, Observations sur les ombres colorees [sic], by HFT Paris 1782 . . . the author derives similar results from similar experiments. I am in the process of creating an extract from this remarkable but sadly little-known work and of presenting it with explanatory notes.'[82]

In the *Historischer Teil* of his *Farbenlehre*, published in 1810, Goethe also stated that it was 'without knowing' H.F.T.'s 'efforts' that he had 'had embarked on the same path' in his 'little essay to the same effect'.[83]

'Von den farbigen Schatten' was certainly much more succinct than H.F.T's text, largely because it employed only four arrangements for directing a large number of paired, white, and coloured light sources on to a white surface. These included day-light, moonlight, direct and filtered candle light, and sunlight reflected off coloured paper. Helpfully, Goethe set out the set-ups he used in a drawing (Figure 2.5), which was intended to be published as an illustration.

The one shown in Figure 2.5 involved projecting two light sources, *a* and *b*, onto the same wall to create two shadows, *eg* and *hf*. The majority of experiments conducted with this arrangement showed that shadows were the same colour as the light illuminating them but not all. In Experiment 6, for instance, Goethe found that the shadow cast by the candle *a* at *eg* was in fact blue, despite being illuminated by (implicitly white) moonlight from *b*.[84] Conversely, in Experiment 12, the shadow

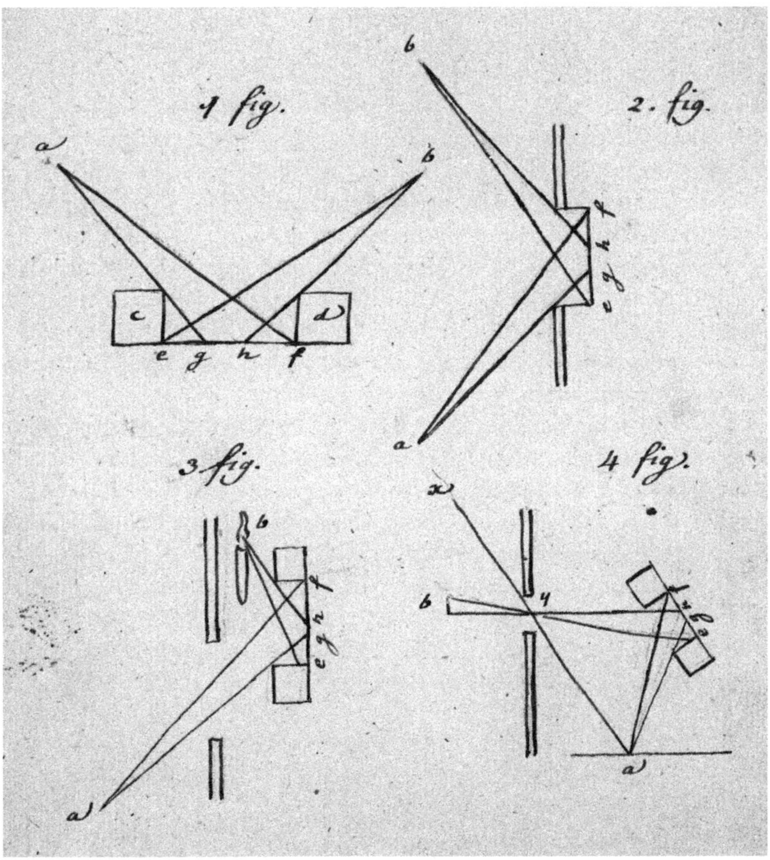

Figure 2.5 Johann Wolfgang von Goethe, original drawing for the essay 'Von den farbigen Schatten', 1792–93. Ink on paper. Weimar, Klassik Stiftung, inv. no. GGz/L.

Source: © Klassik Stiftung Weimar, Bestand Museen.

cast by daylight at *fh* was yellow, although it was lit by candlelight passing through a blue filter.[85]

The closely similar arrangement Goethe employed in Figure 2.5 (second figure) produced no significant anomalies. However, the more complex arrangement shown in Figure 2.5 (third figure), which admitted white sunlight reflected off a white wall at *a* into a chamber where a candle was burning at *b*, did throw up inconsistencies. In Experiment 7, for example, the shadow at *fh* appeared blue, although illuminated only by sunlight.[86] The fourth and most complex arrangement, illustrated in Figure 2.5 (fourth figure), which involved reflecting light from *x* off coloured paper at *a* while admitting daylight from *b*, also generated significant anomalies. Notably, in Experiment 15, the shadow at *eg* lit by direct daylight appeared blue, while the shadow at *fh* illuminated by sunlight reflected off pale blue paper appeared yellow.[87]

In order to explain these departures from the theory of reflection, Goethe – like H.F.T. – argued that the colours of the shadows were determined by the relative brightness, or 'energy', of the two light sources in play.[88] And very likely, it was this claim that prompted Lichtenberg to bring H.F.T.'s argument about the 'proportion of intensity' between the light sources to his attention.[89] Goethe nevertheless drew the original conclusion from his experiments that the weaker of the two light sources always illuminated a blue shadow, and the stronger illuminated a 'yellow', 'yellowish red', or 'yellowish brown' shadow.[90]

Goethe explained this law as a function of how shadow colour, like 'colour' more generally, 'is the result of an effect of light on shadow', drawing his authority from those 'older observers' belonging to the Aristotelian tradition who had explained coloured shadows in terms of a 'mixture of light and darkness'.[91] He relied, in particular, on Athanasius Kircher's argument in the *Ars magna lucis et umbrae* of 1646 that colour is '*lumen opacatum*' or light dimmed.[92] This notion, Goethe argued, is consistent with the idea that a yellow shadow is illuminated by a more powerful light source than the one that casts it. He also argued, by corollary, that a blue shadow could be considered an '*umbra illuminata*' or the darker shadow only weakly illuminated by a light source which is feebler than the one casting it.[93]

Goethe supplemented this argument with an explanation of the pallid coloured shadows Horace-Bénédict de Saussure had seen during a daytime ascent of Mont Blanc in 1787.[94] This came about, Goethe argued, because the 'dark blue' sky that would normally create a 'blue colour in the shadows' did not radiate enough 'light' to do so.[95] What Goethe says here is obscure, and he may not have been satisfied with the general theory of coloured shadows he elaborated in the text as a whole. He was finally so dissatisfied with his essay, at all events, that he refused to publish it in the second volume of the *Beiträge zur Optik*.[96] For this reason, it had no influence on any artist, despite issuing a recommendation to the 'landscape painter' to depict coloured shadows.[97]

The Blue Grotto

The painter August Kopisch made recourse to Aristotelian ideas in his account of the extraordinary blue colour visible inside the sea cave on the island of Capri that he and his colleague Ernst Fries swam into in 1826. They named this cave, once a favourite bathing-place of the Emperor Tiberius, the Blue Grotto.[98] Although the Grotto does not contain coloured shadows, strictly speaking, as its name suggests, it does provide a remarkable spectacle which belongs to the same category of phenomena. This is illustrated in paintings made by Fries in 1826–27 and by Kopisch around 1834. But perhaps it is Heinrich Fried's *The Blue Grotto of Capri* of 1835 (Figure 2.6) that gives the richest impression of its remarkable beauty. Kopisch, who was an accomplished poet, nevertheless delivered a rich evocation of the scene in his essay, *Entdeckung der blauen Grotte auf der insel Capri*, of 1838. This describes:

> [A] massive and deep basin, with a ceiling made of beautiful rocks overgrown by stalactites, the water a billowing sky, whose blue light magically illuminated the ceiling above it. . . . [While] at the entrance the bright daylight shimmered, and like a moonbeam reflected its light over the water.[99]

Kopisch also recounted the 'indescribably enchanting' spectacle that arose 'when the evening sun was shining in the direction of the entrance' and allowed its 'many-pointed vaulting' to be 'displayed in its full glory of colour'.[100] More importantly for present purposes, this sight led him to remark that he could have 'mistaken' the vaulting 'for the blue sky itself, had there not been silver drops falling here and there from the ceiling.'[101] For Kopisch, in other words, the Blue Grotto was an upside-down

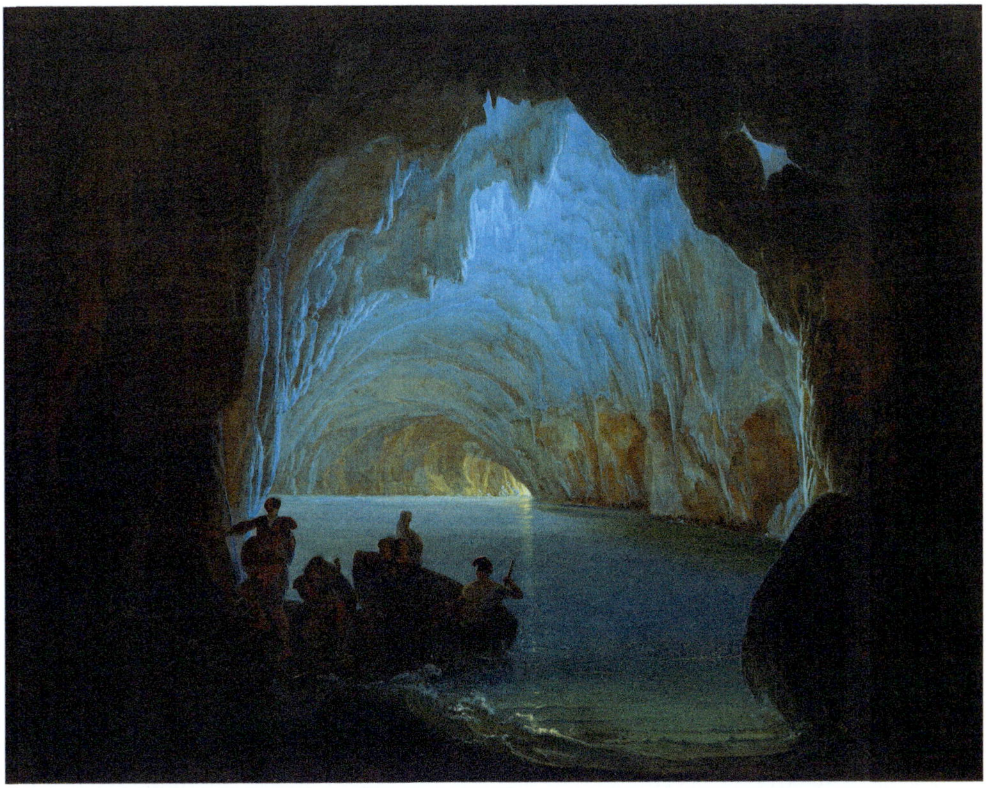

Figure 2.6 Heinrich Fried, *The Blue Grotto of Capri*, 1835. Oil on canvas, 50 × 63 cm. Bremen, Kunsthalle, inv. no. 894–1964/6.

Source: Kunsthalle Bremen /AKG Images.

world in which the sea functioned like the sky to radiate blue light on to any shadowy surfaces beyond the reach of direct sunlight.

Hans Christian Andersen emphasised the same topsy-turvy character of the scene in his novel, *The Improvisatore*, of 1836. His account is delivered through the eyes of his protagonist, Antonio (a figure modelled partly on the Danish painter Thomas Fearnley), who describes his impressions as he stirs out of unconsciousness after being swept into the Blue Grotto by a storm.[102] This reads:

> I lay . . . aloft in the infinitely blue ether which was lighted up around me. . . . I extended my hand down into the strangely shining air below me; it was water into which I thrust it, blue, like burning spirit.[103]

The marvellous effects Kopisch and his colleagues observed were caused by a situation which is not apparent on first acquaintance: that daylight not only enters the Grotto directly through its low mouth, but also indirectly through a large opening below it which is concealed by a bar of rock. Kopisch discovered this fact when investigating why the 'blue' shone 'brightest' from the 'western rock face' above the mouth of the cave, describing how:

> I found that underwater, in the direction of the open sea, there was an enormous opening such that a good diver would be able to swim in and out of the grotto underneath the rock. The rays of light also take this path'.[104]

In an attempt to explain how the Blue Grotto acquired its colour, Kopisch described how the light entering the submerged opening undergoes a form of 'scatter' (first described by Bouguer and discussed in the next chapter). This claim is correct if taken to mean that it is refracted by the sea rather as it would be normally by the sky, so that only its short-wave, 'blue' component emerges in any quantity.[105] (The water in the cave is also bluish owing to its high sulphurous content.) In effect, then, two light sources of different colour temperatures are in operation in the cave: The cool light from the underwater opening and the warmer sunlight that enters through its mouth.[106] The veil of blue light on the Grotto's walls is thus to all intents and purposes a function of the same cause as coloured shadows.

Surprisingly, however, given his scientific interests and his friendship with Alexander von Humboldt, Kopisch chose to explain what he saw in Aristotelian terms.[107] Thus, he argued that 'because the water allows the light to pass into the grotto, while the wider sea serves as a dark background to it, it must appear blue as the illuminated middle, like the air of the sky during the day'.[108] It would appear, then, that the authority of Aristotelian ideas about blueness was remarkably enduring, allowing them to remain a touchstone even for sophisticated investigators of coloured shadow phenomena well into the modern era.

Notes

1. Chapters 28, 104, 155, 158, and 328. The latter is cited in Baxandall 1995, 115; Fiorani 2008, 272–273; and Fiorani 2009, 131–132. On the adaptation of Melzi's manuscript (*Codex Vaticanus Urbinas* 1270) for the *Trattato*, see da Vinci 1939, 5–11; Pedretti 1964, 9–27; and Farago et al. 2018, 1: 213–237.

2. On the obscurity of Leonardo's text, see Sparti 2003 157 n. 65 (citing Fréart de Chambray) and the (unpaginated) preface to da Vinci 1721 (on the 'midnight Darkness' of the text).
3. Farago et al. 2018, 2: 626. See also da Vinci 1540, ff40r–40v; and da Vinci 1651a, 6. This passage is cited in Fiorani 2009, 130. The argument is reiterated in Chapters 104 and 155. See Farago et al. 2018, 2: 680–681 and 713; da Vinci 1540, f63v and f74r; and da Vinci 1651, 27 and 40.
4. Leonardo sometimes envisaged skylight as a reflection of the sun's light and sometimes as a direct emanation from the sky. The first view is elaborated in Chapter 69 of the *Trattato*. See Farago et al. 2018, 2: 654–655; da Vinci 1540, f54r; and da Vinci 1651, 20. The second is developed in Chapter 76, which states that the light of the sky is 'original' (direct) and not 'derivative'. See Farago et al. 2018, 2: 659; da Vinci 1540, f55v; and da Vinci 1651, 22.
5. Farago et al. 2018, 2: 840. See also da Vinci 1540, f148v–149r; da Vinci 1651, 98.
6. This process is set out in Chapter 147, which describes how the 'shadow' of a 'blue' object in a room with 'green' walls 'will be ugly and not the true shadow of such a beautiful blue, because the green reflected on it will discolour it'. See Farago et al. 2018, 2: 709; da Vinci 1540, ff72v–73r; and da Vinci 1651, 39. Leonardo also analysed how objects reflect their colours upon one another in Chapters 77–88. See Farago et al. 2018, 2: 660–669; da Vinci 1540, ff55v–59r; and da Vinci 1651, 18–23. For a discussion of how Leonardo's theory of coloured shadows sits within his more general theory of reflections, see Fiorani 2009, 131 and 140.
7. da Vinci 1540, f148v. For an analysis of this drawing, see Fiorani 2008, 273 and Fiorani 2009, 132–133. For a discussion of the authorship of the engravings in the 1651 editions, see Sparti 2003, 174–180.
8. See Fiorani 2009, 149.
9. For discussion of a similar problem with early botanical illustration, see Ivins 1953, 43–44.
10. For a discussion of the inability of Leonardo's drawing to convey shadow colour, see Fiorani 2009, 133. See also Finley 1967, 366 n. 50 on attempts by Turner and James Sowerby to express colour by variations in an engraved line.
11. It is implicit in this argument that Leonardo already had the theory of coloured shadows by this date.
12. See Fiorani 2009, 125–126.
13. Fiorani 2008, 272. See also Fiorani 2009, 126.
14. Ibid.
15. Fiorani 2008, 271. See also Fiorani 2009, 127.
16. See Fiorani 2008, 271.
17. The 'bluish shadows' Fiorani discerned in the angel's white shirt also looked quite grey to the present author. See Fiorani 2008, 271 and Fiorani 2009, 126.
18. Farago et al. 2018, 2: 626. See also da Vinci 1540, ff40r–40v and da Vinci 1651, 6.
19. Farago et al. 2018, 2: 614. See also da Vinci 1540, f32r and da Vinci 1651, 2.
20. Farago et al. 2018, 2: 614. See also da Vinci 1540, f32r and da Vinci 1651, 2.
21. For philosophical discussions of the grain of experience, see Evans 1982, 229; Peacocke 1992, 67–68 and 83–84; McDowell 1994, 56–60; and Cohen 2001, 114–116. For the role played by categories in memories of colour, see Foster 2011, 689–690 and Raffman 1995, 294–297.
22. See Hering 1964, 6–12. See also Katz 1935, 163–165 on 'memory colours' and for the argument that 'language' causes us to 'exaggerate' colours so that we remember them as being more 'saturated' than they are, which inclines the painter to select a colour of an 'exaggerated saturation' to match what he sees. For a neuroscientific confirmation of Hering's thesis, see Vurro et al. 2013. On 'real colour', see Merleau-Ponty 1996, 304–305 and Kelly 2005, 82–87 and 90–98, which describe it as an 'absent norm'.
23. Baxandall 1995, 115. See also Kemp 1999, 131, for the argument that Leonardo conceived of blue shadows as 'reflections' of the sky.
24. For a close analysis of Leonardo's indebtedness to Aristotle for this theory, see Bell 1993, 102–111.
25. Farago et al. 2018, 2: 711. See also da Vinci 1540, ff73r–73v and da Vinci 1651, 39.
26. Farago et al. 2018, 2: 701. See also da Vinci 1540, ff70v–71r and da Vinci 1651, 35–36.
27. See Bell 1993, 110. On the authorship of this text, see Gottschalk 1964. On the theory it elaborates, see Hoeppe 2007, 28–30. On Leonardo's access to the manuscript, see Pedretti

1964, 56–57 n. 62. See also Bell 1993, 101–103 and 111–115, for Leonardo's knowledge of, and borrowings from, *De Coloribus*.

28. Aristotle 2000, 19. See also Aristotle 1957, 233–235 for the corollary argument that the sun is red when its light is seen through a turbid medium.
29. Farago et al. 2018, 2; 690. See also da Vinci 1540, f67r and da Vinci 1651, 31. For a discussion of Leonardo's ideas about the production of blue by a process of superposition, see Bell 1993, 105–109.
30. Farago et al. 2018, 2: 819. See also da Vinci 1540, f140v and da Vinci 1651, 89. Cf. Chapter 340. See Farago et al. 2018, 2: 855; da Vinci 1540, f53v; and da Vinci 1651, 103.
31. See Kuhn 2012, 148.
32. Barrow 1669, 86. This translation by H.C. Fay is from Barrow 1987, 153.
33. Barrow 1669, 86.
34. Ibid.
35. Ibid.
36. The image was subsequently republished in Barrow 1674, opposite 92.
37. For this explanation, see Barris 1999, which claims in addition that the diagram implies that the paper takes on the 'reddish cast of the candle flame' in the segments X-A and β-Y, where it is illuminated by this light source on its own.
38. On Guericke's theory of coloured shadows, see Baxandall 1995, 116 and 176, n. 33 and Land 1977, 126.
39. Guericke 1672, 142.
40. Newton 1704, 97–98.
41. See Shank 2008.
42. On Gautier d'Agoty and his anti-Newtonianism, see Lowengard 2006 ('Industry and Ideas: Jacques-Fabien Gautier, or Gautier d'Agoty'); Margócsy 2014, 194–199; and Lavezzi 1999.
43. Gautier 1750–51, 2: 70.
44. See ibid. 14–18 for this theory, which is neo-Aristotelian in maintaining that light is nothing other than the 'impulsion' or 'compression' of 'fiery globules' emitted by the sun among the particles of 'transparent' bodies such as air or water. Light thus construed is subject to different degrees of 'compression' and variations in the 'speed' of its 'vibrations'. Rays of a 'simple and uniform pressure' are 'white', but rays whose vibration is 'composite' as a result of encountering solid objects are variously coloured. And the blue of the sky is produced when light runs into the 'black and opaque air' through a process akin to refraction involving scatter, which Gautier called 'lateral reflection'. For another theory of the production of colour from black and white involving 'vibrations', see Castel 1740, 13–14, 48, and 96–97. On Gautier's debt to Castel, see Margócsy 2014, 195 and 264, n. 105. See also Castel 1740, 232 for the argument that: 'In general shadows are coloured, even yellow, or a little green by virtue of being mixed a little with blue, which makes the true black colour'.
45. Gautier 1750–51, 2: 19. See also ibid., 170 for the claim that blue is a 'pure & transparent light, through which we see shadow'. For Castel's ideas about the involvement of 'movement' and 'repose' in the mixture of light and dark that produces colour, see Castel 1740, 48.
46. Gautier 1750–51, 2: 49–50. See also ibid., 33, on the role played by this form of interposition in producing the colours of spectra.
47. Ibid., 50. See also ibid. 33, which explains that 'lateral reflection or refraction occasions the interposition of light over shadow' and makes the sky appear blue when white light from the sun is reflected back off the darkness of space, as enfeebled light whose vibrations have been slowed down.
48. Ibid., 70.
49. Ibid.
50. Ibid.
51. Gautier 1752, 1: 108.
52. See H.F.T. 1782, 132, which mentions the three-volume second (1750–1751) edition of the *Croa-Génésie*.
53. See ibid., 88 for H.F.T.'s claim that he was the first to investigate 'artificial light', neither Buffon nor Millot having examined shadows of this kind.
54. Ibid., 130–131.

55. Ibid., 132.
56. Ibid., 91–92.
57. Ibid. See also ibid., 101–102 for a description of the blue shadows produced by two candles suspended in birdcages; ibid., 133–134 for the argument that the colour of a shadow produced by two candles will change when one of them is moved; and ibid., 179–180 for the claim that a colour appears on one part of a white surface when this was less strongly illuminated than the parts around it.
58. Ibid., 133; citing Gautier 1750–51, 2: 60 and 40.
59. See H.F.T. 1782, 138, citing Gautier 1750–51, 2: 215 on how 'the particles of bodies . . . reflect or refract light under different modulations', thereby producing both 'fixed and accidental colours'; and H.F.T. 1782, 133–135, citing Gautier 1750–51, 2: 19 on shadow as the 'repose' of light.
60. H.F.T. 1782, 133–135, citing Gautier 1750–1751, 2: 49–50.
61. H.F.T. 1782, 135–136, citing Gautier 1750–1751, 2: 50.
62. H.F.T. 1782, 138–139.
63. Ibid., 139.
64. Ibid., 140–141.
65. Ibid., 141. Ford Madox Brown rendered the coloured shadows produced under similar conditions in *Waiting* of 1854–55. In this the shadows in the white robe of the baby are red where they are illuminated by the fire and blue in appearance where they are lit by the lamp (although the paint here is physically grey). On this work, see Brown 1981, 76 and 78.
66. H.F.T. 1782, 97 and 141.
67. Ibid., 172–173.
68. Ibid., 173.
69. Ibid.
70. Ibid., 177. See also ibid., 180–181 and 183.
71. Ibid., 185–186.
72. Ibid., 186–190.
73. Ibid., 183.
74. Ibid., 183–184.
75. Ibid., 213.
76. Ibid., 197.
77. Joost 2002, 301–302. For Goethe's reply to Lichtenberg in October 1793, see Goethe 1887–1912, 10: 115–121. For other letters mentioning coloured shadows, see Goethe ibid., 9: 274, 311, and 314 and 10: 95, 66, 104, 111. On Goethe's essay, see Ott 1977, 5 and Schöller 2017, 18.
78. The essay was originally intended to form part of the second volume of the *Beiträge zur Optik*, published in 1792. See Sepper 1988, 88 and Boyle 2000, 98, which state that the essay was written 'largely in 1792 though not finished until 1793'.
79. See Goethe 1897, 102, 104, 107, 110, 119, and 122. See also Sepper 1988, 89–90.
80. Goethe 1897, 102–104.
81. Ibid., 302, citing Gehler 1790, 2: 826.
82. See Goethe 1887–1912, 10: 126.
83. Goethe 1810, 2: 612.
84. Goethe 1897, 107.
85. Ibid., 112.
86. Ibid, 108 and 105–106.
87. Ibid., 114.
88. Ibid., 116. See also ibid., 108–110 and 114.
89. See Joost et al. 2002, 302, citing H.F.T. 1782, 197. On Lichtenberg's strategy in this letter, see Joost 2001, 31. See also Goethe 1840, 363–364 for Eastlake's note to the effect that Goethe cited H.F.T. in the 'Historical Part' of the *Farbenlehre* on how: 'the colour of shadows is as much owing to the light that causes them as to that which more faintly illumines them'.
90. Goethe 1897, 110.
91. Ibid., 120.
92. Ibid. Kircher does not use this exact phrase.

93. Ibid. See also Boyle 2000, 163.
94. Regarding Goethe's familiarity with Saussure's ideas, see Goethe 1887–1912, 9: 274, for a letter to Johann Heinrich Voigt of June or July 1791, in which he mentioned how, upon reading in the 'March issue of the Journal de Physique on page 199 the description of M. de Saussure's blue-measure (Cyanometre) [sic], I remembered all the observations on the colour blue I have had occasion to make, and I considered anew the theory I had formed about it.' The article mentioning the cyanometer Saussure designed for measuring the sky's colour was Saussure 1791. On this, see also Saussure 1779–96, 4: 198 and 290–292 and Hoeppe 2007, 110–111.
95. Goethe 1897, 123. See Saussure 1779–96, 4: 296 for the author's similar belief that this dark colour is a result of how the 'black void of space' mixes with the 'natural colour of the air'.
96. The manuscript remained unpublished until 1897. It is now in the Klassik Stiftung, Weimar. Inventory number, GSA 26, Li. 13, S. I f. (*Schatten*). 1792–93.
97. Goethe 1897, 123–124.
98. On Tiberius, see Houston 1985. On the Grotto's name, see Kopisch 1838, 192–193. For the context, see Kittelmann et al. 2016, 125 and the catalogue entry by Stefano Tumidei in Cavina 2001, 251.
99. Kopisch 1838, 181. See also the translation by Veronica Galbiati in Kopisch 2016.
100. Kopisch 1838, 195.
101. Ibid., 196.
102. On this passage and Andersen's use of Fearnley as a model for Antonio, see the catalogue entry by Stefano Tumidei in Cavina 2001, 251. On Andersen and Fearnley, see Andersen 2005, 149 and 204–205. See also Cavina 2001, 256, for an illustration of Fearnley's *Terrace of Capucins at Sorrento* of 1834, which depicts a blue shadow, like several of the artist's paintings.
103. Andersen 1845, 1: 156–159. See also ibid., 2: 329–330.
104. Kopisch 1838, 196.
105. On 'scatter' in the sea, see Minnaert 1954, 314–315.
106. Kopisch 1838, 197.
107. On Kopisch and Humboldt, see Petersdorff 2016, 50 and Kittelmann et al. 2016, 154, and 238.
108. Kopisch 1838, 196–197.

3 Physics

Conceptions of Reflection

Leonardo's argument that blue shadows are coloured by the light of the sky was not lost on subsequent scholars. Newtonians like Nollet, for example, glossed over it by stating that a surface 'visibly assumes the colour of the azure light of the sky' where 'there is a shadow'.[1] But while this statement implied that a blue shadow appeared blue because it reflected the light of the sky, it did not actually say so.

The medical scientist and natural historian Johann Daniel Major had come closer to doing so, however, in a description of an experiment concerning the colour of snow which he wrote in 1672 and which was translated into French in 1761.[2] This reads:

> We have often observed the light reflected by snow in [a] darkened chamber . . . by passing the rays of light returned from areas covered with snow through a lens. When the snow that returned the ray of light was illuminated directly by the rays of the sun, the image painted on the wall of the darkened chamber was white; but when the snow that returned the ray of light was in shadow, the image painted on the wall was as blue as sapphire – the blue colour of the Sky (perhaps) being painted on the surface of the snow as in a mirror, when this surface is not exposed directly to the rays of the sun.[3]

Clearly, Major used the word 'reflected' in this passage to denote the process whereby a shadow returns the light of the sky to the spectator's eye, even likening it to the action of a mirror.

The situation is nevertheless complicated by the fact that other physicists used the word 'reflect' and its cognates in a second way to refer to the process underlying the latter, namely the one whereby the sky casts its light into a shadow. In his translation of Chapter 28 of the *Trattato*, for example, Fréart de Chambray maintained that 'shadows' are coloured because the 'blue of the sky reflects and gives its colour to the parts [of a scene] it sees' – even though Leonardo's Italian made no mention of reflection at all here.[4] Similarly, the edition of Jacques Ozanam's *Recréations mathématiques et physiques* published by Charles-Antoine Jombert in 1778 declared that:

> [T]he shadow cast by bodies exposed to the sun is . . . itself lit by the whole part of the sky the luminous body does not occupy. This part of the sky being . . . blue, the shadow is broken by blue or azure rays, and must appear this colour. . . . [It] is nothing more than a shadow mixed with the reflection of a blue body, and consequently it must participate in this colour.[5]

Jombert even added the Leonardesque rider to his remarks on: 'It is precisely as, in painting, reflections are tinged with the colour of surrounding bodies'.[6]

What complicates matters further is that physicists considered skylight to be a reflection of sunlight and hence used 'reflect' in yet another way to describe this formative process. Bouguer, for example, described how the sky is blue because it 'reflects' light from the sun.[7] And defining this process in terms of scatter, he mentioned how the 'blue rays of light [are] reflected more easily' than red and also characterised blue skylight as the 'reflection produced by the particles of the air'.[8] This conception of reflection also informs Bouguer's explanation of how shadows are coloured by blue light from the sky. As he explained:

> [T]he shadows of the morning and evening take on a very blue hue, and that of a candle produces nearly the same effect, when it takes the place of the sun, which has not yet risen, but is just about to appear. This phenomenon is caused by the aerial colour of the atmosphere which illuminates these shadows, and in which the blue rays dominate: [T]hey reflect obliquely in quantity, while the red rays . . . cannot change the shadow, because they do not reflect, or reflect much less.[9]

Bouguer's notion that scatter was a form of reflection probably owed something to Newton's *Opticks* of 1704, which maintained that the 'azure Colour of the Skies' was 'reflected' by the 'small parcels' their 'vapours' formed.[10] Newton's conception was in any event recapitulated by Thomas Melvill in his paper, 'Observations on Light and Colours', first delivered in 1752 in which he claimed that the atmosphere 'reflects more blue rays than what go to the formation of perfect white'.[11]

The ambiguity attaching to words whose root is 'reflect' also affected statements by artists. Among these, Valenciennes, who was the first artist to write extensively about coloured shadows after Leonardo, used the word 'reflection' in two ways. Thus, on the one hand, he used the word in the *Elémens de perspective pratique* in its familiar sense to refer to the way that blue shadows reflect skylight into the spectators' eye. On the other hand, he used it to describe how light from the sky travels into shadows, when he stated that 'Shadows are always illuminated' since 'they almost always owe their colour to the blue reflection from the part of the celestial vault opposite the light'.[12] In one passage of the *Elémens de perspective pratique*, he even employed both usages in close succession, when describing how it is possible to see 'reflections' of a 'bluish tint' on the sides of objects opposite the 'light' at dusk, which 'owe their existence' to the 'reflection of the azure celestial vault'.[13]

Valenciennes was not alone in running these two usages together. Indeed, Beguelin had done so earlier in his essay, 'Sur les ombres colorées' of 1767, when he wrote:

> [T]he source of those blue rays which . . . appear . . . in the shadows of morning and evening . . . is to be found quite naturally in the pure air, which itself appears blue to us . . . [because it] reflects the rays which excite the sensation of this colour in preference to all others. All objects within reach of receiving the direct rays of the sun, are at the same time exposed to receiving a greater or less quantity of the rays that the air reflects . . . [so] it is not surprising that the part that is in the shade can reflect some towards us.[14]

Beguelin was nevertheless careful here to distinguish the process whereby 'reflected' skylight coloured shadows from the one wherein reflections from adjacent objects did so.[15] The same cannot be said, however, of Michael Huber's 1775 translation of Christian von Hagedorn's *Betrachtungen über die Mahlerey* of 1762. Notably, this added a footnote to Beguelin's essay in a chapter on 'reflections', which not only cited Leonardo's ideas about the colours objects reflected on one another but also described the 'mixed reflections' visible in the skin tones of Adriaen van der Werff's *Diana Discovers Calisto's Sin* of 1704.[16]

Given that it was traditional in the eighteenth century to regard coloured shadows as reflections of one sort or another, it is no surprise that the term resurfaced in the writings of nineteenth-century artists.[17] Among these was Delacroix, who tended to use it to refer to the light shadows emitted. He stated in his journal for May 1830, for example, that 'All planes in shadow' must 'each have its own particular reflection; for example, all planes which face the sky' must be 'bluish'.[18] Then, in March 1831, he mentioned 'The shadow of white objects very strongly reflected in blue'.[19] And in January 1857, he stated that 'There are no shadows strictly speaking: there are only reflections'.[20] Although Delacroix did not mention Leonardo's *Trattato* in this context, it may have coloured his thoughts, however, since he had considered writing a commentary on the text which would 'embroider upon its dryness'.[21]

The notion that coloured shadows have the colour they did because they reflect the sky persisted into the Impressionist era. In 1876, for instance, Edmond Duranty stated in *La Nouvelle peinture* that with the Impressionists, 'painters have, for the first time, understood and reproduced' how nature 'reflects' the 'hue of the vault that arches over the earth'.[22] Duranty also referred to the light of the sky itself as a 'reflection'.[23] And in 1905, Paul Cézanne harked back to the same antiquated sense of 'reflection' when he told the artists R.P. Rivière and Jacques Schnerb that Claude Monet 'had known to colour shadows' because they were 'areas which, deprived of direct sunlight, only receive a reflection from the sky'.[24]

Cézanne's statement rehearses what was by then received wisdom. He may have learned about reflection much earlier, however, from the only book on colour he owned, *De la lumière et de la couleur chez les grands maîtres anciens* of 1865, by the painter and colourman, Jean-Désiré Régnier.[25] This maintained that 'coloured shadows' were illuminated by the sky and hence that 'The shaded parts of objects are most often illuminated to some extent' by the 'radiation of light in the ambient air'.[26]

How Cézanne became aware of Régnier's book is not known. One possibility is that it was brought to his attention by Camille Pissarro, with whom he discussed 'theories' in the 1870s according to Lucien Pissarro, and who was the 'theorist' of the Impressionist movement if Renoir is to be believed.[27] And it is likely that Pissarro first became aware of the colour of shadows from reading Valenciennes, since in December 1883, he sent a copy of the *Elémens de perspective pratique* to Lucien, with a letter telling him it 'is by the famous Valenciennes. It is ancient, but it is still the best and the most practical [book]. Try to take account of the principles that serve as its foundation'.[28]

The theory of reflection elaborated in these books manifests itself in Monet's *The Magpie* of 1868–99 (Figure 3.1).[29] The fact that the snow is untrammelled, or pristine, in the scene depicted indicates it is most likely a painting of the early morning; but it would be naive to suggest it simply records what Monet saw out of doors for a short period over several days. Even in the 1860s, the artist worked on paintings in the studio for long periods, during which time he inevitably made recourse to memory and to theories of what he had seen. This mixed pattern of work is evident from the paint layers in the magpie itself, which suggest the following sequence of events. First,

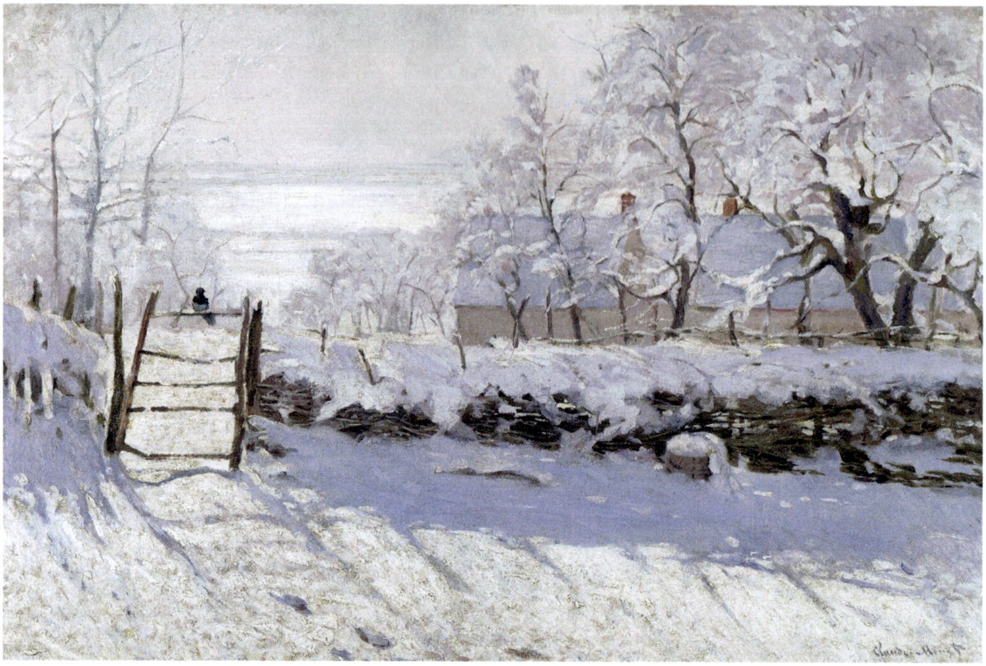

Figure 3.1 Claude Monet, *The Magpie*, 1868–69. Oil on canvas, 89 × 130 cm. Paris, Musée d'Orsay, inv. no. RF 1984 164.

Source: Musée d'Orsay/Bridgeman Images.

Monet blocked in the shadow of what is now the tail with a thin layer of blue paint. He then added a thick layer of white paint depicting snow over this, with the result that parts of the tail were submerged beneath it. Subsequently, Monet superimposed slightly darker blue paint corresponding to the right side of the magpie's body over, and to the right of, the initial layer representing the tail. He evidently did so some considerable time after the first campaign since the texture of the already dry paint in the tail shows through the new layer.

More importantly, Monet not only elaborated this area after the fact, but he also did so incorrectly. That is, the orientation of the shadow is that of the bird facing to the right, whereas the bird in the painting is facing the other way. It is the wrong way round, in other words, just like a mirror-image. It is just possible, then, that Monet made this error because he over-generalised from the theory that blue shadows reflected, or mirrored, the blue of the sky.

Whatever the case in this instance, the evidence is conclusive that the idea that shadows were coloured by processes involving reflection caught the imagination of painters after the upsurge of scientific interest in Leonardo's ideas. What is more, given the demonstrable connection between their awareness of the theory and the appearance of coloured shadows in their work, it seems to have made these elusive effects noteworthy and noticeable.

Dawn and Dusk

One of the great advantages of the physical theory of reflection is, oddly enough, that it restated Leonardo's ideas on the blue colour of the shadows visible at dawn and dusk in terms which were straightforward enough to appeal to painters. By corollary, it would seem that Leonardo's ideas remained too obscure without reinterpretation to have significant effect. So, although it is tempting to think that they lay behind Henry Mauperché's *People on the Steps of a Destroyed Palace* of *c.* 1645 (Figure 3.2), one of the earliest paintings besides Leonardo's of a sunset scene in which contrast gives the shadows a faintly blue colour, there is no evidence that the artist had access to any of the manuscripts of the *Trattato* circulating in France at the time.[30]

A possible exception to this rule is Valenciennes's *The Ancient City of Agrigentum* of 1787 (Figure 3.3). In this, coloured shadows are visible alongside patches of warm sunlight in the left foreground and in the white walls of the architecture in the distance. This distribution recalls Leonardo's description of coloured shadows in Chapter 328 of the *Trattato*, although this could be fortuitous. However, it would seem that Valenciennes was aware of this text since he advised the painter in the *Elémens de perspective pratique* to 'acquire the science of colour' through 'seeking the cause of the effects we see' and to 'establish the principles that give the general

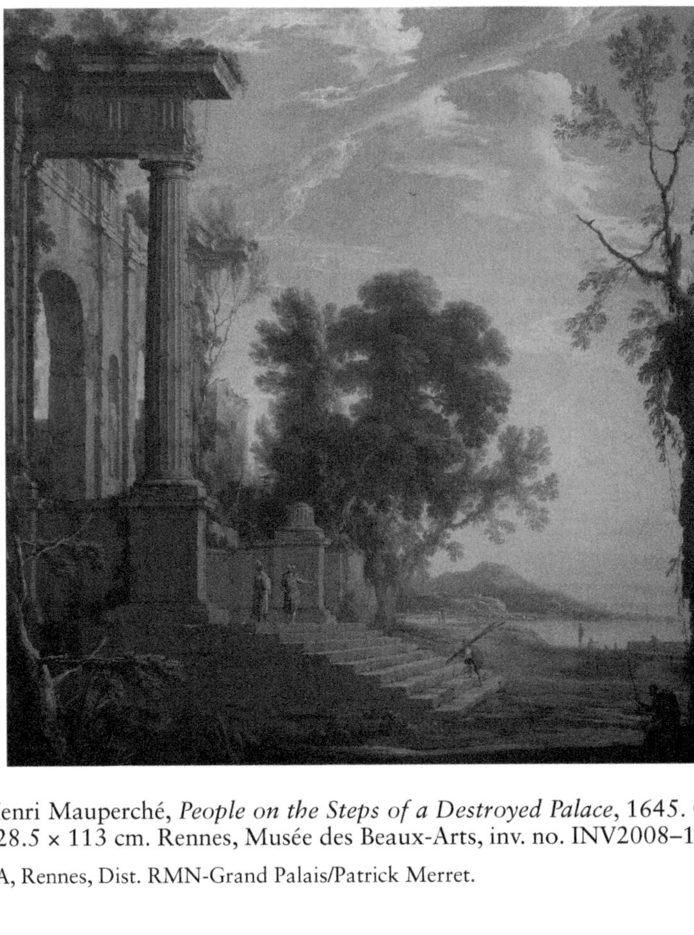

Figure 3.2 Henri Mauperché, *People on the Steps of a Destroyed Palace*, 1645. Oil on canvas, 128.5 × 113 cm. Rennes, Musée des Beaux-Arts, inv. no. INV2008–1–1.

Source: © MBA, Rennes, Dist. RMN-Grand Palais/Patrick Merret.

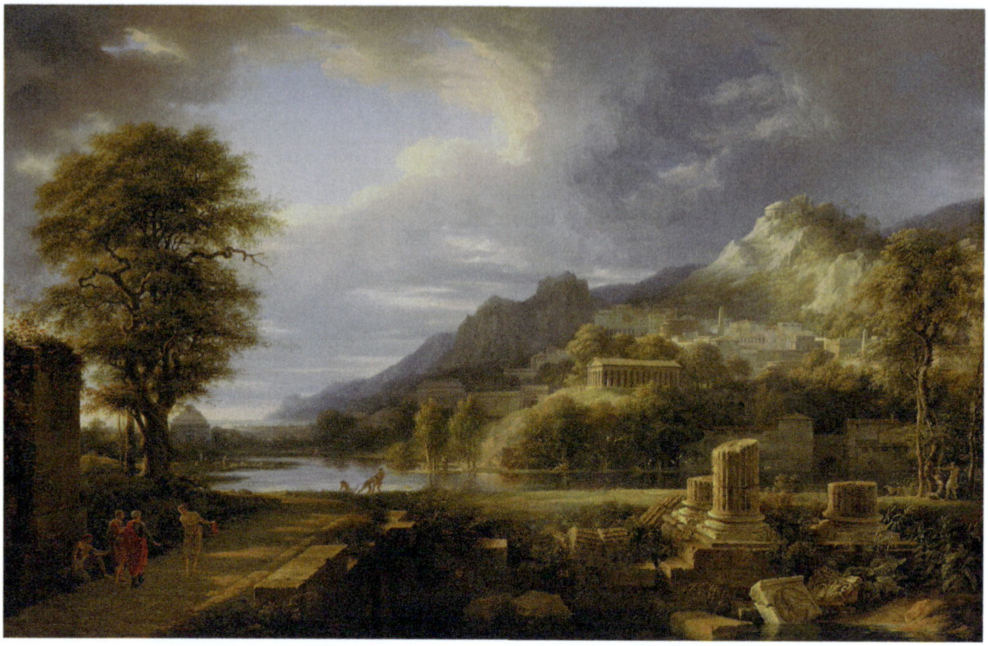

Figure 3.3 Pierre-Henri de Valenciennes, *The Ancient City of Agrigentum*, 1779. Oil on canvas, 110 × 164 cm. Paris, Musée du Louvre, inv. no. MNR48.

Source: Musée du Louvre/Bridgeman Images.

rule' for representing particular 'effects', very much as Leonardo had.[31] Moreover, Valenciennes recommended the artist to commit the 'effects' he saw to 'memory' when pressed for time, just as Leonardo did.[32]

The Ancient City of Agrigentum is, moreover, a synthetic work which observes these principles. At all events, no painted studies for the painting survive, even though Valenciennes recommended the painter to observe the landscape for just 'half an hour' at 'sunrise' and 'sunset' and to make oil sketches 'in the open air', which recorded the appearance of 'nature under the vault of the sky, receiving direct light, reflections, and shadows from the atmosphere'.[33] Similarly, neither the pencil sketch Valenciennes made of the site when he first visited Agrigento in 1779, nor the notes he made about it in his diary contain anything about colour.[34] It can therefore be classified as a 'historical landscape' of the kind Valenciennes urged the artist to imbue with 'ideal beauty' by using his 'imagination', rather than a 'landscape portrait' in which the 'faithful representation of nature' or 'resemblance' was the paramount consideration.[35]

Leonardo's impact on Valenciennes notwithstanding, it is possible that more recent science played a part in determining the painting's treatment of coloured shadows. In the *Elémens de perspective pratique*, for example, Valenciennes acknowledged his familiarity with the 'long, learned dissertations' written by 'physicists' on the 'colour of shadows', although he claimed he did not 'understand' these, as they were 'beyond the reach of artists who only study Nature'.[36]

Physicists' writings about coloured shadows spread to Germany in the later eighteenth century, as they were translated and discussed in scientific journals.[37] And along with them so did Valenciennes's, whose treatise was translated into German in 1803. It is possible that Runge knew something of these sources. He was, in any case, one of the first German artists to articulate the idea that the sky coloured shadows blue at dusk in a short text of *c.* 1805–07, 'Sonnen-Untergang im Walde'. Describing the colours he would use to paint a 'sunset in a wood', Runge stated that 'where there is shadow' and 'an air-reflection', he would use 'slightly bluish [paint], with an airy tone to it'.[38]

Although the accompanying line drawing gives no indication of what he meant, Caspar David Friedrich, whom Runge met in 1801 before joining him in Dresden between 1802 and 1806, depicted a closely similar scene in *Hill and Ploughed Field near Dresden* of *c.* 1824–25 (Figure 3.4).[39] In this, the rising sun is below the horizon, behind the hill in the distance, but its warm light can be seen through the screen of trees in the foreground, rather as in Runge's sketch. Here, too, both near and distant objects are tinged with light reflected by the blue sky. On the trunks of the foreground trees especially, small blobs of blue paint have been superimposed over earlier blue smears.

Valenciennes's ideas probably had an impact on the Danish artist, Christoffer Wilhelm Eckersberg, although opinion is divided over the matter.[40] It is known, for

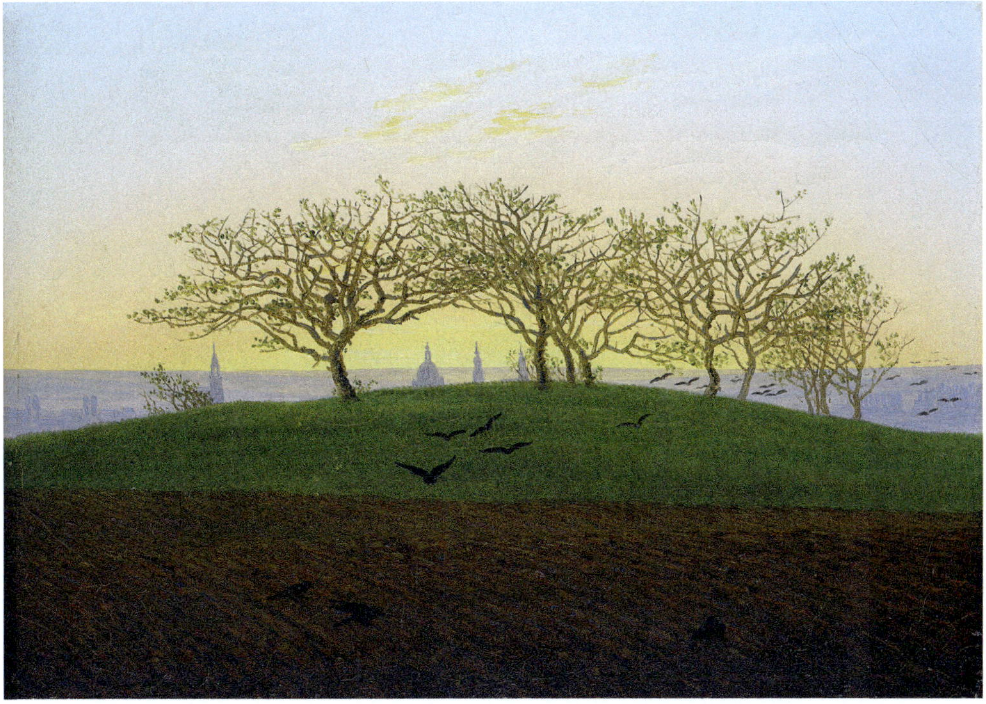

Figure 3.4 Caspar David Friedrich, *Hills and Ploughed Field Near Dresden*, 1824–25. Oil on canvas, 22.2 × 30.4 cm. Hamburg, Hamburger Kunsthalle, inv. no. HK-1055.

Source: Hamburger Kunsthalle/Bridgeman Images.

example, that Eckersberg's teacher at the Royal Academy of Fine Arts in Copenhagen from 1803 to 1809, Nicolai Abilgaard, owned a copy of the German translation of the *Elémens de perspective pratique*.[41] And it is possible that Eckersberg met Valenciennes, or heard of his ideas, while working in David's studio in Paris between 1811 and 1812.[42] When Eckersberg was Professor at the Royal Danish Academy of Fine Arts in 1831, moreover, he introduced the practice of painting in the open air as part of the curriculum.[43] The Danish artist's compositional arrangements and his meticulous observation of the patterns of shadow specific to particular times of day also suggest a debt to Valenciennes.[44] It makes sense, therefore, to think that the unusually saturated blue shadow visible in the foreground of Eckersberg's *A View From the Château of Meudon near Paris* of 1813 (Figure 3.5), a 'scene' he described in a letter of October 1812 as 'especially beautiful in the evening sunshine', owes something to the French painter's ideas.[45] So, too, in all probability, do the touches of light blue paint on the top face of the block at the front of the pile of stones to the right of the young woman carrying a basket.

In England, painters first became interested in coloured shadows through their direct acquaintance with physics. Or at least this seems to be the case with Joseph

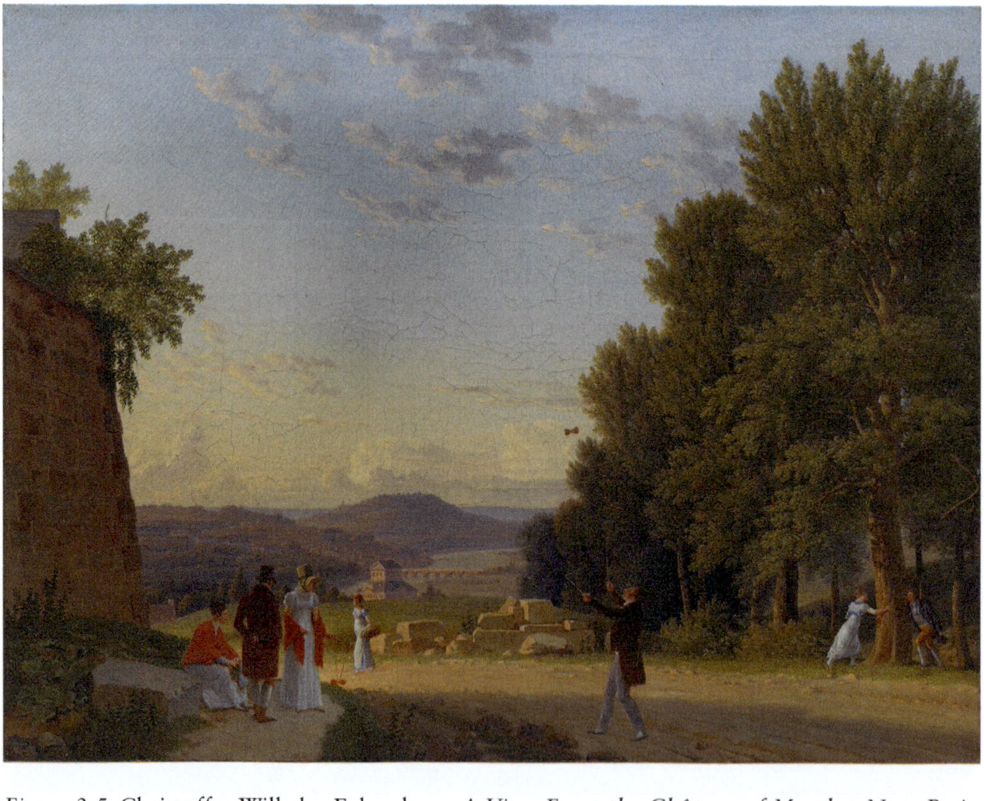

Figure 3.5 Christoffer Wilhelm Eckersberg, *A View From the Château of Meudon Near Paris*, 1813. Oil on canvas, 55.5 × 71 cm. Copenhagen, Statens Museum for Kunst, inv. no. KMS1623.

Source: SMK/Jakob Skou-Hansen.

Wright of Derby, who depicted coloured shadows in the middle distance of his sunset scene, *A View of Cicero's Villa, Pozzuoli, Near Naples* of *c.* 1789 (Figure 3.6).[46] Being scientifically minded, Wright may have read the long summary that Priestley made of Buffon's, Beguelin's, and Melvill's arguments about coloured shadows in *The History and Present State of Discoveries Relating to Vision, Light, and Colours* of 1772.[47] Wright may even have met Priestley when he visited Bath in the mid-1770s, while the painter was living there.[48]

Wright was certainly acquainted with several members of the Lunar Society – an association of natural philosophers and industrialists, which met in and around Birmingham from around 1766 and which included Priestley from around 1780.[49] During the 1780s, more particularly, Wright was a visitor to the Lunar cénacles held at Matthew Bolton's Soho mansion in Birmingham (and at Josiah Wedgewood's house in Etruria in Stoke-on-Trent).[50] The Lunar Man, Erasmus Darwin, had also been Wright's doctor since 1767 while the painter was resident in Lichfield.[51] Darwin then moved to Derby in 1782.[52] So, even if Wright did not know about Priestley's work on coloured shadows directly, he would probably have known that it was mentioned by Darwin in *The Loves of the Plants*, first published in 1789.[53]

The extent of the renown enjoyed by Priestley's ideas is indicated by the fact that the novelist and essayist, Maria Edgeworth, whose father Richard was a Lunar Man,

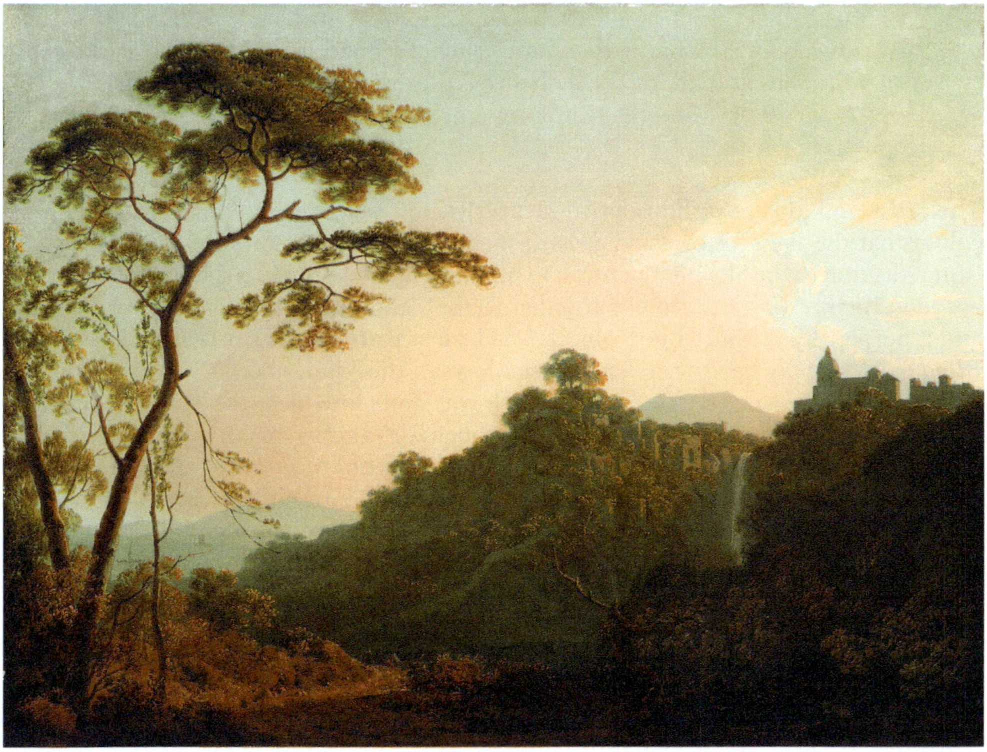

Figure 3.6 Joseph Wright, *A View of Cicero's Villa, Pozzuoli, Near Naples*, *c.* 1789. Oil on canvas, 59.5 × 77.3 cm. Schorr Collection, on long-term loan to the Henry Barber Trust, The Barber Institute of Fine Arts, University of Birmingham.

referred to them in her stories.[54] In 'Waste Not, Want Not', for example, published in 1800, one of the characters pauses to admire 'coloured shadows', while the reader is directed by a footnote to 'Priestley's *History of Vision*, chapter on coloured shadows'.[55] Later, in *Harry and Lucy Concluded* of 1825, Edgeworth devoted a short episode to coloured shadows which cites Priestley's account without mentioning its author's name. This delightful passage begins with Harry looking at the landscape outside as it is projected by a *camera obscura* on to piece of paper, whereupon Lucy exclaims:

> 'I know what you are thinking of, Harry, of the coloured shadows we used to see on the wall of our room at home.' She darted out of the room, and returned with a quarto volume. 'Now, Harry,' cried she, 'for blue shadows and green! black shadows and red! I will read you all about them.' She sat down and read – 'It is rather remarkable, that so curious an appearance as this of blue shadows should pass unnoticed near a century, and should then be hit upon by mere accident. Buffon, as he was busy about something else, observed that the shadows of trees which fell upon a white wall were green.' Here Lucy read a curious but long account of his observations upon green and indigo shadows, seen in different circumstances, ending with, 'any person may see a blue shadow, if he will only hold his finger before a piece of white paper at sunrise or sunset.'
>
> Harry said he should like to try this.
>
> 'Is it not all very entertaining?' asked Lucy.
>
> 'Very,' said Harry; 'but is there not some explanation given? What is the cause of the different colours of these shadows?' 'There is a great deal about it in this book,' replied Lucy, 'and the history of a great many experiments, which different people have tried.'[56]

The artist and amateur philosopher, Henry Richter, subsequently discussed the 'blue' colour that the 'light' of the 'sky' gave to shadows and the importance of admitting the 'soft reflection of the Sky into pictures' in his pamphlet, *Day-Light* of 1817.[57] By his own account, Richter was attuned in particular to the 'delightful' effect that the 'evening Sun and a purple Sky produce upon objects'.[58] His lost painting, *Christ Giving Sight to the Blind* of 1812, may have even depicted shadows of this kind, since an anonymous critic recorded how the 'sun of glory' in this work was quite low, 'near the head of Jesus'.[59]

Richter exhorted artists to work in the 'open air' and to 'make genuine *studies of light and colour*, taken faithfully from *Nature* itself, *out of doors*'.[60] According to his son's recollections, moreover, he 'sketched much in oil' himself.[61] Richter's enthusiasm for painting in daylight owed something to his admiration for the philosopher Kant and his emphasis on the 'phenomenon' *per se*.[62] He envisaged his own art, therefore, as a corrective to Classical painting based on the art of antiquity, made by artists who 'judged what they saw' and 'boldly copied in their works THE CONCEPTIONS THEY HAD FORMED IN THEIR MINDS'.[63] While such commitments led Richter to advocate the 'necessity of attending to the natural operation of Light', he nonetheless insisted on the importance of 'reflecting upon the causes of those phenomena which the painter cannot always have by the side of his easel'.[64] Despite his open-air rhetoric, therefore, Richter appears to have apprehended coloured shadows in much the same way as Leonardo and Valenciennes; that is, under the guise they assume when assimilated to a theory of their origins.

William Blake knew Richter.[65] He also read Darwin's *Botanic Garden*, the second volume of which comprised *The Loves of the Plants*, before making etchings for it in 1791.[66] It is likely, therefore, that some of the works Blake produced over a long period, in which a setting sun casts blue shadows while imparting a warm glow to the lights, were predicated on the theory of reflection these texts expounded. Blake's tempera painting, *Satan Smiting Job With Sore Boils* of *c.* 1826 (Figure 3.7), is one candidate. It should be noted, however, that in his copy of *The Works of Sir Joshua Reynolds*, Blake annotated the author's observation that the 'illuminated parts of objects are in nature of a warmer tint than those that are in the shade', with the remark: 'Shade is always Cold'.[67] Taken together with the fact that Blake used the same colour scheme in many works, some depicting underwater scenes, it seems that it was to some extent formulaic or synthetic – much like *houding* or *cangiante* colour – and sometimes lacked a perceptual rationale.[68]

Later, the artist Harry Willson cited Leonardo's *Trattato* several times in his book, *The Use of a Box of Colours* of 1842.[69] Here, he also argued that shadows were coloured both indirectly by light reflected from nearby objects and directly by light sources of one kind or another.[70] Willson even cited Leonardo's claim that 'Shadows produced by the redness of the setting sun, will be blue; from the reflexes of that part of the air not illumined by its rays.'[71] And by way of an explanation, he continued: 'If

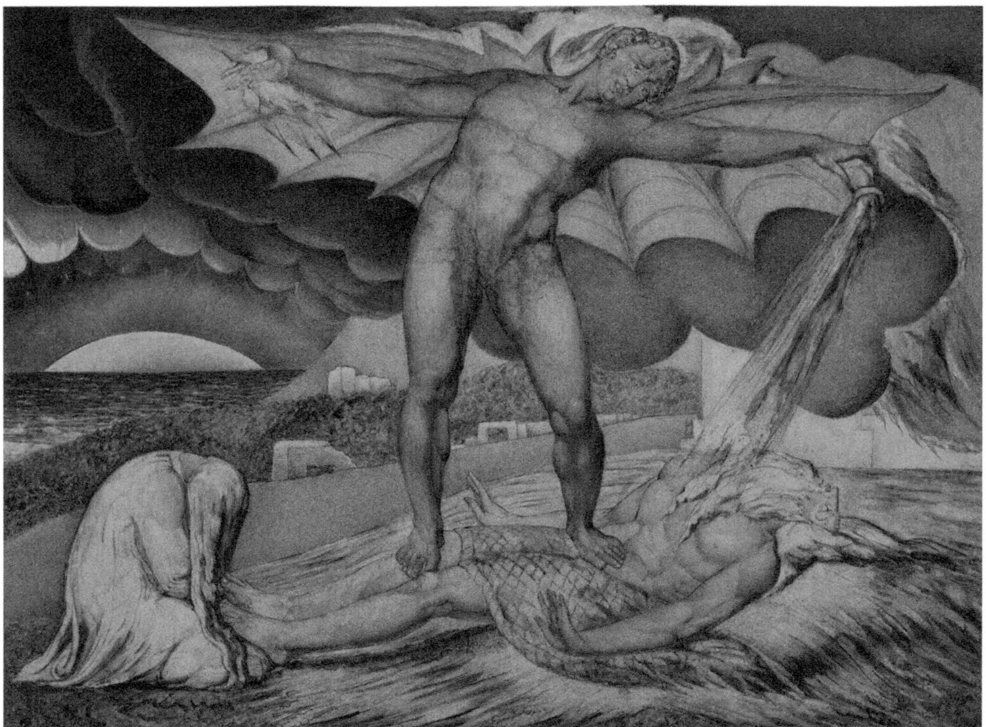

Figure 3.7 William Blake, *Satan Smiting Job With Sore Boils*, *c.* 1826. Ink and tempera on mahogany, 32.6 × 43.2 cm. London, Tate Britain, ref. no. N03340.

Source: © Tate.

the sunbeams burst out' those 'objects in the light will participate in *their* colour from reflection', while those 'parts not included in the range of rays, remain the colour of the *air*'.[72] It is broadly consistent with these remarks, then, that Willson recommended that the 'masses of light should be of warm colours, yellow or red, supported by blue or grey in the shadows'.[73]

Leonardo's work was also known to more famous painters. F.G. Stephens stated in 1860, for instance, that William Holman Hunt not only relied on 'his own observations' in *A Converted British Family Sheltering a Christian Missionary From the Persecution of the Druids* of 1849–50 (Figure 3.8), but also implemented the 'peculiar effect' previously 'hinted at by Leonardo da Vinci' in giving their 'true colour to sun-shadows' and in deepening these 'shadows to pure blue where he found them to be so'.[74] The saturation of the blue shadows cast by the megaliths in the background of the painting and the peculiarly vibrant blue–violet shadow on the rear wall of the interior of the hut suggest a time of day not long after dawn or just before dusk, which could indicate Hunt's indebtedness to the several chapters of the *Tratatto* concerned with these times of day. Hunt did not even mention Leonardo, however, when he

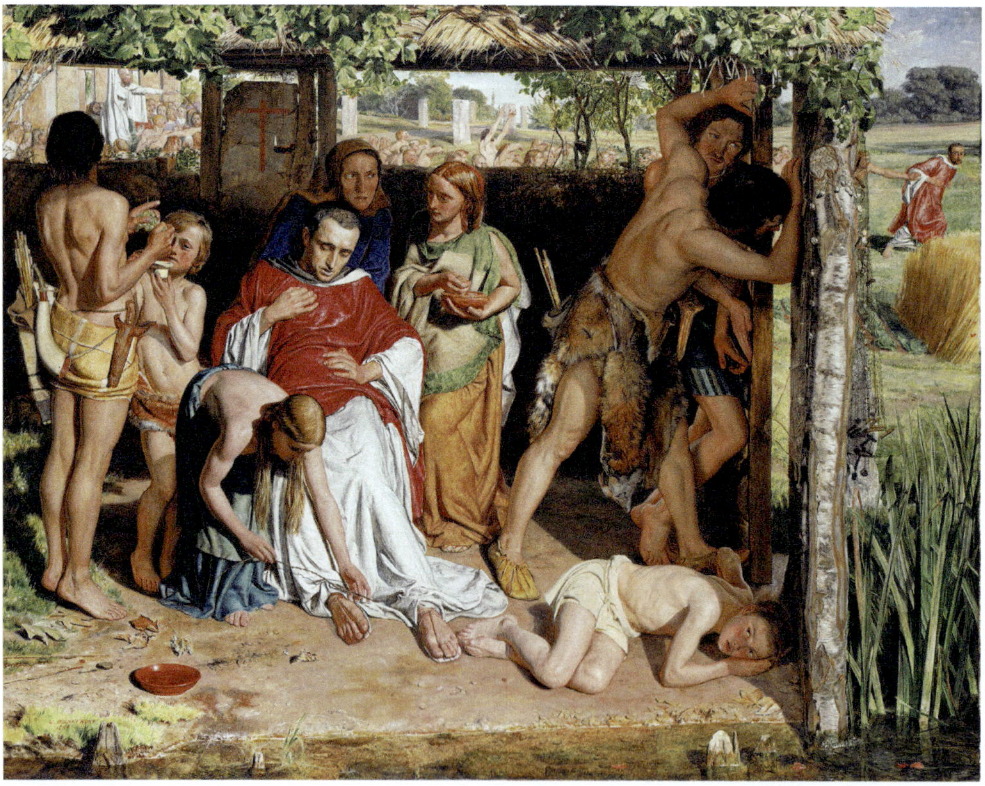

Figure 3.8 William Holman Hunt, *A Converted British Family Sheltering a Christian Missionary From the Persecution of the Druids*, 1849–50. Oil on canvas, 111 × 141 cm. Oxford, Ashmolean Museum, bequeathed by Mrs Thomas Combe, 1893, acc. no. WA1894.1.

Source: Ashmolean Museum, University of Oxford/Bridgeman Images.

recollected the 'open-air effect', he and his colleagues 'had been surprised to discover while searching for the truth before Nature herself', nor when he recalled that they 'registered prismatic hues' because they 'found that each terrestrial feature mirrored the blue sky and the tints of its neighbouring creations'.[75]

One difficulty with any interpretation of Hunt's intentions in *A Converted British Family* is that the light in the scene is somewhat brighter than it is around dawn or dusk. This can be explained, however, by the fact that synthesising effects belonging to different times of day was unavoidable for painters of the period because the pigments available to them did not permit a large canvas to be covered quickly. Indeed, by his own account, Hunt worked 'long' and 'hard' on the painting in Lea Marshes on sections of the 'background', 'foreground', and 'hut', 'with no studio' over the course of a 'month' during August 1849 before continuing work on the painting in the studio, where he painted his models 'as the sun shone into [his] window in the early morning'.[76]

Sky-Shadows

Although Leonardo saw that the surfaces of white objects were coloured 'red' or 'reddish' by the setting sun, he did not conceive of them as lying in shadow. Similarly, Beguelin observed that 'rays of sunlight' at sunset produced a 'red' or 'reddish-yellow' border below the blue shadow his finger cast on a piece of paper, but he did not describe this as a shadow, either.[77] The distinction implicit in their writings, between a surface lit by the sun and a shadow lit by the sky and, is a little incongruous. It is no surprise, then, that as the notion that a shadow could be illuminated gained ground, so did the idea that areas illuminated by the weak sunlight of dawn or dusk lay in shadow.

H.F.T. was one of the first to articulate this belief when he described 'coloured shadows' of a 'beautiful yellow', which were 'illuminated by the sun'.[78] It is perhaps unsurprising that the first painter to mention sky-shadows, Carl Gustav Carus, was also a scientist.[79] In a letter of October 1823 published in his *Neun Briefe über Landscaftsmalerei* of 1831, he detailed in his characteristically precise but poetic language how 'one evening', as the 'sun dipped into the diaphanous veils that lined the horizon' and rendered these a 'delicate rose red', the same 'rose red reflected from the shadow side of every object'.[80]

Willson described sky-shadows in *The Use of a Box of Colours* when he mentioned how 'Every object receiving the light of the sun, receives likewise the general light, producing two shadows'.[81] Here he also mentioned 'warm shadows' which 'partake of the colour of the light'.[82] Hunt's *A Converted British Family* depicts very distinct, orangey sky-shadows of this kind in the flesh of a boy cowering in the bottom right corner, sometimes next to more conventional blue–violet shadows. They were undoubtedly designed to represent the warm sunlight that not only struck the figure directly, but was also reflected upwards from the soil and the water.[83] It nevertheless stretches credulity to believe Hunt could actually have seen orange colours of such saturation.

Willson analysed a more complex kind of shadow in a passage stating that when the 'horizon is tinged with red by the rays of the setting sun, the distant shadows, being blue or azure, mingling with the red, produces [sic] purple'.[84] And obligingly, Willson illustrated composite coloured shadows of this kind in one of the plates he made for his book, *A View in Belgium* (Figure 3.9). Purple shadows can also be seen in Cézanne's painting, *Millstone in the Park of the Château Noir* of 1892–94 (Figure 3.10), particularly on the blocks of white limestone. They, too, are composite

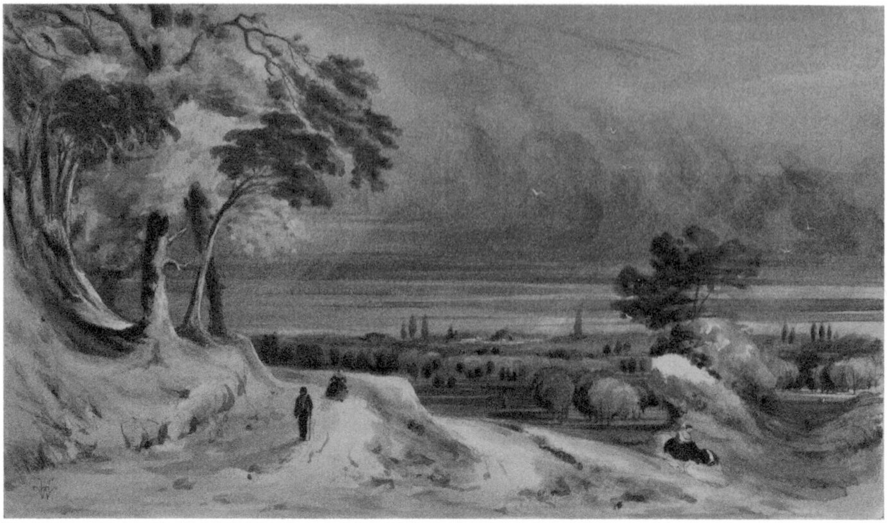

Figure 3.9 *A View in Belgium*. Plate VIII from Harry Willson, *The Use of a Box of Colours* (London: Tilt and Bogue, 1842).

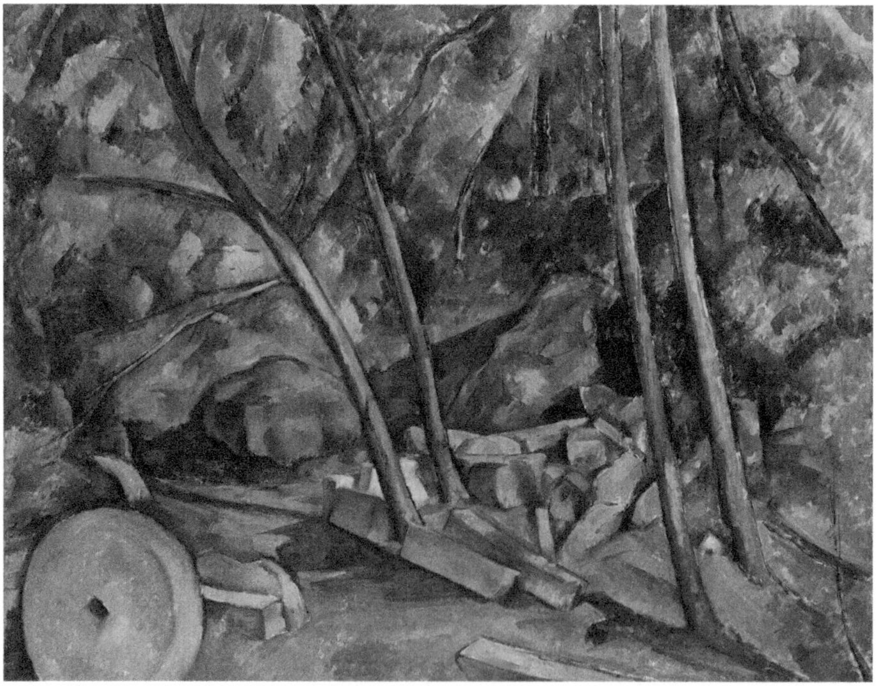

Figure 3.10 Paul Cézanne, *Millstone in the Park of the Château Noir*, 1892–94. Oil on canvas, 73 × 92 cm. Philadelphia, Philadelphia Museum of Art, The Mr and Mrs Carroll S. Tyson Jr Collection, 1963, acc. no. 1963–116–4.

Source: Philadelphia Museum of Art/Bridgeman Images.

shadows, although in this case they are blue shadows into which warm light has been reflected from the soil below. Cézanne could have been prompted to notice them by a passage in Régnier's *De la lumière et de la couleur*, which describes how 'along with the radiation of light in the air there is also a luminous radiation from the mass of the sunlit ground upon which things seen rest'.[85]

Shadows of this kind are illuminated only indirectly by the sun, but Régnier described sky-shadows which are lit by direct sunlight as well. Although the passage in which he did so is ambivalent, it might be translated like this:

> When the air is very dilated, the light of the sun is very intense; the sky is all the more blue as we approach the equator. When, of an evening, or in the morning, the sun is roughly forty degrees above the horizon, every object it illuminates will cast blue shadows on the bright ground, and shadows of a broken colour leaning more towards orange than any other hue.[86]

It is unclear from Régnier's text whether he believed such shadows could appear in the 'North'.[87] Notwithstanding, it is possible that his ideas were of interest to Monet, since warm colours can be discerned on the sunlit sides of several of the grain stacks he painted at dawn and dusk over the winter of 1890–91 – even though he told Gustave Geffroy in a letter of October 1890 that the 'sun sets so quickly that I can't keep up with it'.[88] In *Grainstack, Snow Effect, Morning* of 1891 (Figure 3.11), for example,

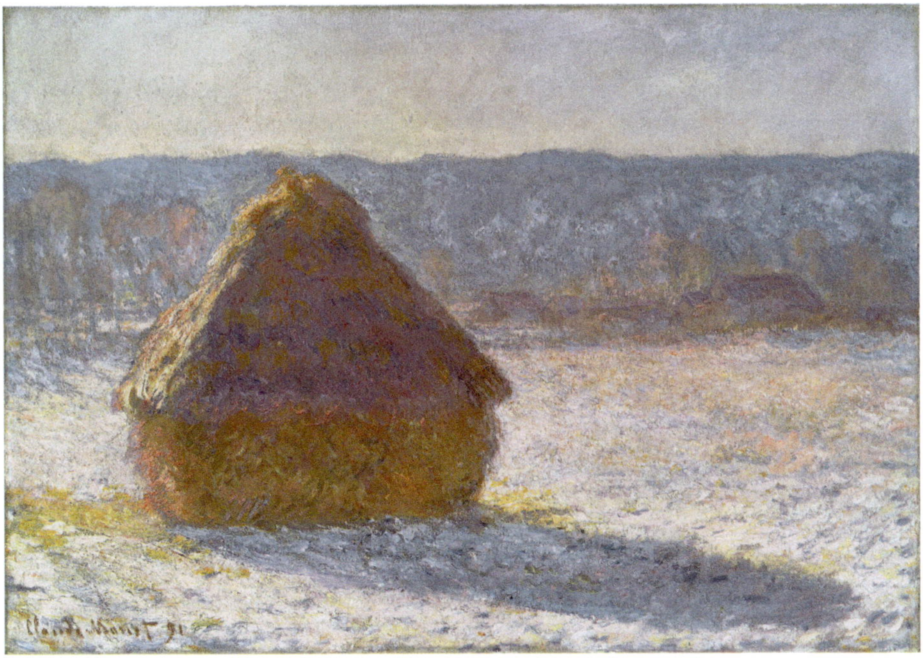

Figure 3.11 Claude Monet, *Grainstack, Snow Effect, Morning*, 1891. Oil on canvas, 65.4 × 92.4 cm. Boston, Museum of Fine Arts, gift of Miss Aimee and Rosamond Lamb in memory of Mr. and Mrs. Horatio Appleton Lamb, acc. no. 1970.253.

Source: MFA/Bridgeman Images.

there is a pronounced sky-shadow on the left of the stack. This distribution seems correct since it makes sense to think that they occur in areas that the bluest portion of the sky – which is at around 95° from the sun towards dawn and dusk – does not reach.[89] Smaller areas of warm shadow can be seen on the periphery of the large blue shadow to the right of the stack. These, too, seem to be veridical, since they have a distribution similar to the yellow fringe Beguelin observed below the blue shadow cast by his finger.

Another possible source for Monet's and Cézanne's interest in sky-shadows is Ernst Brücke's *Die Physiologie der Farben für die Zwecke der Kunstgewerbe* of 1866, which was translated into French in the same year. Although this did not use the word 'shadow' to describe these effects, it did characterise all lit only by the sun or by the sky, indifferently, as partially illuminated surfaces. It thus described how, for example:

> [T]he limestone escarpments of the Alps, on a beautiful day under a blue sky, their grey appears yellow, where it is directly illuminated by the sun, but bluish and cold opposite, in the places illuminated only by the light reflected by the firmament. When the sun descends towards the horizon, this difference becomes more pronounced, the colour of the rocks exposed to the sun turns reddish yellow, the lighting growing warmer as it were towards the evening.[90]

Brücke also described the unusual purple shadows that can only be seen in mountainous situations on snow. The concerned effect occurs when light from the setting sun illuminates airborne ice or snow to produce a band of red light in the sky, which in turn illuminates shadows in the snow, which are already lit by the blue sky. Or, as Brücke put it:

> If a valley is open to the west, the rays of the setting sun, coloured red by their passage through the more dense lower layers of the atmosphere, can still illuminate [its] walls of rock. This lighting reaches its maximum warmth at a few minutes before sunset, causing the phenomenon known to Swiss tourists under the name of fire of the Alps (*Alpenghühen*). If the sky is red at this moment, the cool parts of the scene are also warmed [by the sun], but their hue is more purple.[91]

Brücke directed his advice to painters; but the conditions under which this effect manifests itself made it very difficult to capture the effect before the advent of colour photography.

Sunlight

Although it was a central tenet of physical theory that a clear sky reflected blue light into shadows from sunrise to sunset, investigators were divided over whether blue shadows were actually visible at all times of day.

Leonardo seemed to acknowledge that that they were, at least sometimes. In Chapter 104 of the *Trattato*, for example, he argued that 'Shadow on white, when seen in sunlight and open air, tends towards blue.'[92] And in a passage of his notebooks not included in the *Trattato*, he stated that: 'If you see a woman dressed in white in the country, that part of her which is exposed to the sun will be bright in colour' and the part 'exposed to the luminous air' tend 'towards blue since the air is blue'.[93] In Chapter 155, however, Leonardo added a caveat to these observations, arguing that the 'shadows' of a 'white' object 'will be blue' wherever it 'faces' the 'air' and can 'take on colour' from it, only if it is located 'at high elevation'.[94] And while he did not specify here that this effect only occurred at particular times of day, he did do so in Chapter 158, where he stated that 'white objects' on 'snow-covered mountains' will exhibit 'bluish shadows' only when the 'sun sets on the horizon' and it 'appears to be on fire'.[95]

Gerard de Lairesse was slightly less permissive in the pronouncements he made in *Het Groot Schilderboek* of 1707. In one place, he argued that the shadows of objects were not coloured by the 'Blue of the Sky' at all, 'when the Sun or light Clouds shine not on them', the sky being unable to 'to impart this Colour to the Objects' on account of its 'vast Remoteness or Height'.[96] Elsewhere, Lairesse described how shadows can appear 'reddish Grey' in September, 'about 2 or 3 in the Afternoon, when the Sun is strongest' and the sky has a 'clear blue Colour'.[97] Despite insisting, however, that such shadows do not unite 'with the Blue as in common Light, as some imagine', he conceded that they could become 'gradually a little more Violet' under the influence of skylight.[98]

Buffon, for his part, unequivocally denied that coloured shadows were visible in daytime, stating in his paper of 1743 that that they did not appear at 'midday and all other times of day' apart from dawn and dusk.[99] He gave no explanation for this observation, but Beguelin did in his paper of 1769. Noting that 'sunlight is weakest at sunrise and sunset, and increases in strength as the sun approaches its high point at midday', he argued that 'The weaker the sun's light, the more that the contrast between the shaded and illuminated part of a white wall is softened' and the 'better' the 'colour' can appear 'in the shaded part'.[100] This happens, he explained, because although the 'part of the wall which is in the shade must actually receive blue rays all day long', the 'brightness of the daylight obscures the sensation of these rays in us'.[101] Beguelin's argument has been confirmed by modern science, which has demonstrated that when the sun is strong, brightness contrast effectively darkens shadow colour to the point of rendering it almost black, or colourless.[102]

According to a letter Millot sent Buffon, who published it in the 1774 re-edition of his essay on accidental colours, 'shadows are coloured' not only at 'dawn and dusk', but also at 'midday' under some circumstances.[103] Millot described, more particularly, how 'very beautiful blue' shadows can be seen indoors in daylight on a sheet of white paper, when this is illuminated by a combination of diffuse daylight and the light of the sun as it breaks through the 'clouds'.[104] Later, H.F.T. picked up on this argument to support a claim that blue shadows only occur when there is a degree of 'equilibrium' between the two light sources producing them.[105] Although H.F.T. argued that this normally occurred only 'at sunrise and sunset' when the 'sun has lost its vivacity', he explained that this had in fact been the case with Millot's shadows, since they had appeared when a 'great part of the solar rays were intercepted by the clouds'.[106]

The same line of reasoning informs H.F.T.'s claim that 'coloured shadows' can be seen outdoors, even 'when the Sun was very high above the horizon' in the right conditions.[107] He recorded, for example, how:

> On the 20th of February, between two at three o'clock in the afternoon, I saw shadows coloured a bright violet at some moments. The sky was strewn with clouds, some of which were thin and so transparent, that in passing over the sun, they merely weakened its light a little without intercepting it entirely. In these moments when the sun's rays were weakened, the Shadows on the branches of a tree attached to a white wall which faced the Sun directly, took on a violet hue.[108]

Valenciennes's pronouncements on the visibility of coloured shadows in sunlight are somewhat equivocal. As already mentioned, he argued in the *Elémens de perspective pratique* that shadows are always lit by the light from the sky, from which he concluded that 'an object in the open air' consequently 'produces a shadow' which 'will be bluish', adding that 'If the body' is 'of a light tone, it will be all the more blue'.[109] He did concede, however, that the 'very lengthy Shadows of the rising and setting sun' are 'bluer than at other times of day'.[110] And it is significant that the only painting of his to represent coloured shadows depicts a sunset scene.

As Goethe noted, Saussure had raised doubts about the visibility of coloured shadows in daylight in his *Voyages dans les alpes* of 1796. Here, more especially, the explorer recalled how 'despite the intensity of the blue colour of the air in these high regions, the shadows projected by the sun never appeared dark blue' in the daytime, but were instead 'pale' when 'blue', 'pale violet', 'yellow', or outright 'colourless' or 'black'.[111] It is ironic, then, that the hand-coloured etchings that Chrétien de Méchel produced in 1790 to illustrate Saussure's ascent depict saturated blue shadows in broad daylight – even though the scientist took a close interest in them when they depicted him in action, as for instance does *Voyage de Mr. Saussure a la Cime du Mont-Blanc au mois d'Août MDCCLXXXVII. II^de^. Planche* (Figure 3.12).[112]

English scientists were, by and large, more convinced of the visibility of blue shadows under daylight than their European counterparts. Melvill, for example, stated in his 'Observations on Light and Colours' of 1752 that the 'blueish colour of sky-light may be seen on bodies illuminated by it' on a 'clear cloudless day', when the 'shadow' cast on a sheet of 'paper' by the 'sun's direct rays' appears 'remarkably blueish'.[113]

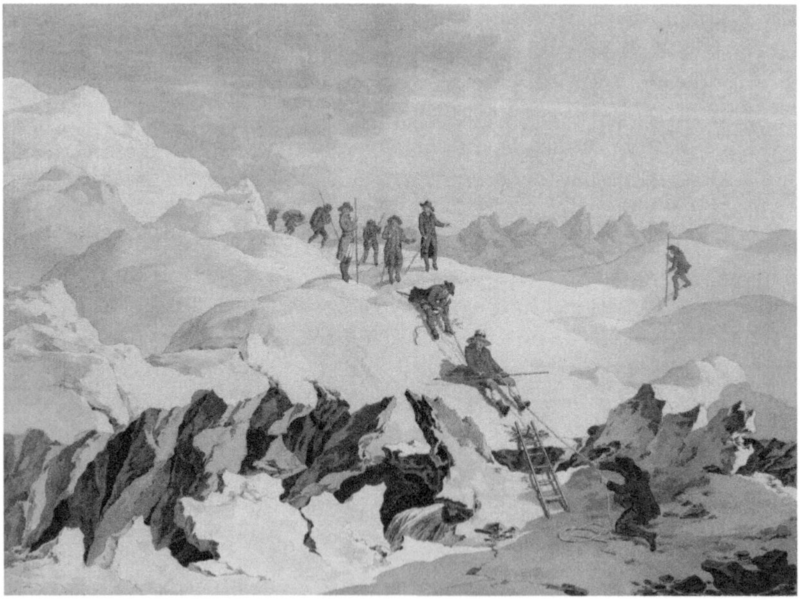

Figure 3.12 Chrétien de Méchel after Marquard Wocher, *Voyage de Mr. Saussure à la Cime du Mont-Blanc au Mois d'Août MDCCLXXXVII. II^{de}. Planche*, 1790. Etching with hand colouring, 34.4 × 35.7 cm. London, British Museum, museum no. 1958, 0712.2821.

Source: © Trustees of the British Museum.

Darwin, who believed that the 'particles of air' were 'themselves blue', stated in *The Loves of the Plants* that in consequence of this a 'blue shadow may be seen at all times of the day', although he did add that they did so 'much more beautifully in the mornings and evenings, or by means of a candle in the middle of the day'.[114] And subsequently, Henry Brougham remarked in his article, 'Experiments and Observations on the Inflection, Reflection, and Colours of Light' of 1796, that he had observed 'blue shadows' at 'all times' on a sheet of paper 'illuminated by the sky, and any other light'.[115] Brougham also described an experiment demonstrating how this happened, which involved lighting an object with the 'blue, and violet, and red' rays refracted by a prism to colour the shadows on one side of it.[116]

Perhaps because English scientists were so sanguine about the possibility of seeing coloured shadows at all times of day, nineteenth-century English artists had few reservations about their visibility under daylight. In his book of that name, Richter explained his ideas on the subject through the ghost of Teniers, who states that 'Past twelve o'clock, and day-light shines down perpendicularly'.[117] Richter helpfully illustrated this claim in a diagram (Figure 3.13), which shows how a 'HORIZONTAL

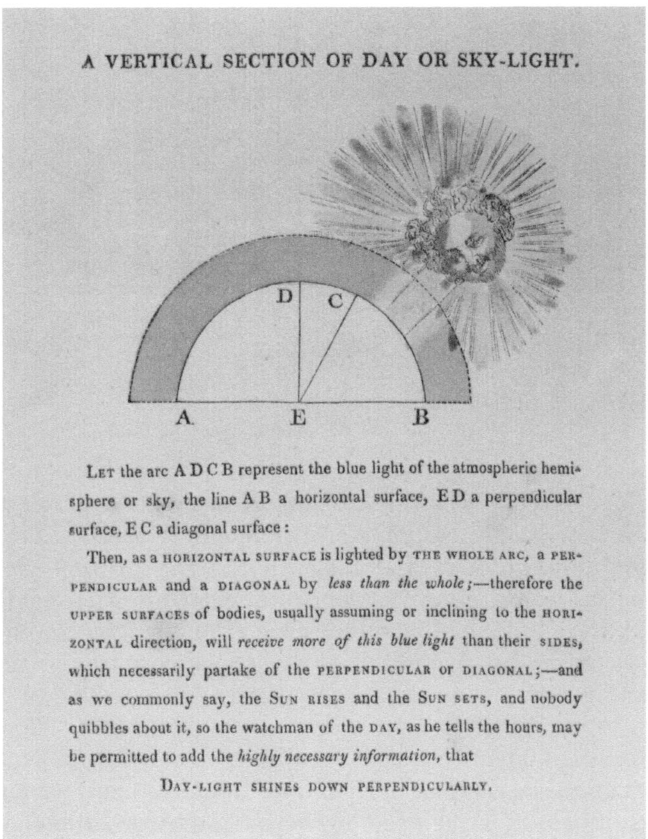

Figure 3.13 A Vertical Section of Day or Sky-Light. Frontispiece to Henry Richter, *Day-Light* (London: Ackermann, 1817).

Source: Author.

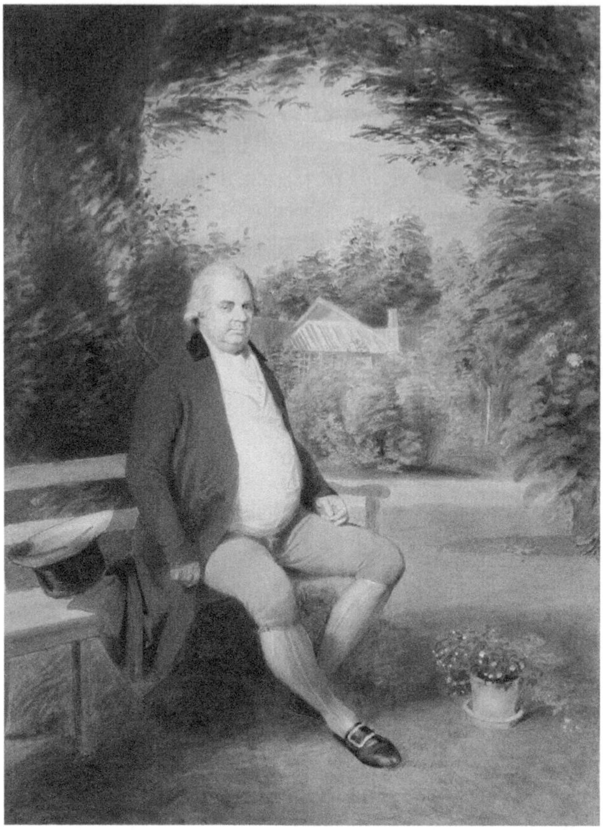

Figure 3.14 Henry Richter, *Portrait of a Gentleman*, *c.* 1830. Watercolour, 48 × 35 cm. Private collection.

Source: Sworders Fine Art Auctioneers.

SURFACE is lighted by THE WHOLE ARC' of the sky, with the result that the 'UPPER SURFACES of bodies, usually assuming or inclining to the HORIZONTAL direction, will *receive more of this blue light* than their SIDES'. He also applied this principle in paintings such as his watercolour of a gentleman seated in a garden (Figure 3.14) in which the sitter casts a blue shadow on the ground to his right. Despite its superficial plausibility, however, the basic geometry of Richter's theory was hopelessly mistaken, since the sun is at its bluest in an arc approximately 65° from the sun when it is at its height.[118]

Richter's book was known to artists. In a letter of January 1879, Samuel Palmer acknowledged that his ideas about the action of 'cool sub-light' from the 'sky' on the 'upper shadows of objects in the open air' was a 'discovery', although he believed it 'savoured of truism'.[119] It has been suggested that Richter's ideas affected Brown's treatment of the blue shadows that fall on horizontal surfaces in *Chaucer at the Court of Edward III* of 1847–51 (Figure 3.15) – which are particularly pronounced on the top and steps of the font and on the floor at the jester's feet.[120] This does not rule out

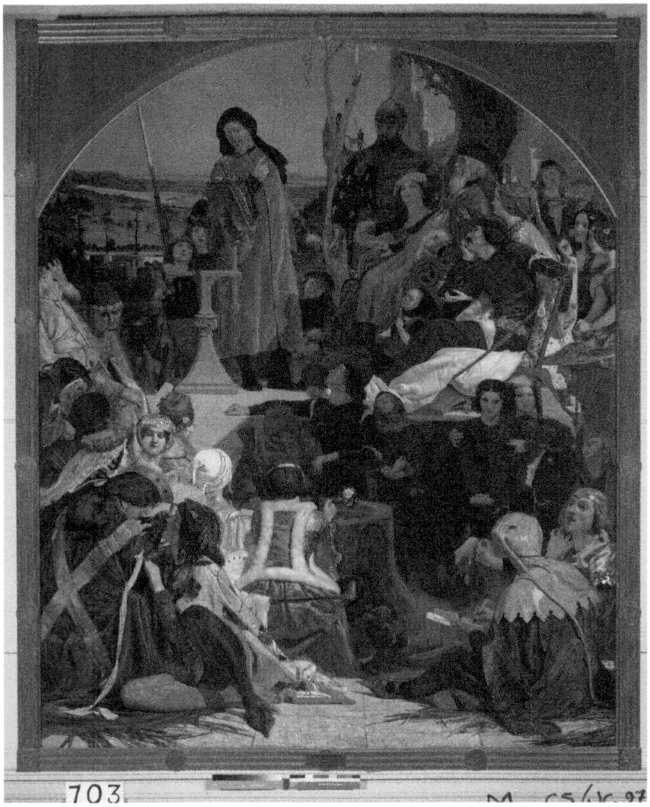

Figure 3.15 Ford Madox Brown, *Chaucer at the Court of Edward III*, 1847–51. Oil on canvas, 372 × 296 cm. Sydney, Art Gallery of New South Wales, acc. no. 703.

Source: Art Gallery NSW/Bridgeman Images.

the possibility that Brown did actually see shadows of this kind, however. He stated, for example, in the catalogue *The Exhibition of Work* he produced in 1865 that his intention in this painting was to treat 'Sunlight not too bright, such as it is pleasant to sit in out of doors'.[121] His diary for 1849 also shows that he did do much of the work on the painting in his studio under 'sun light'.[122] Notwithstanding, it would appear that Brown was averse to relying on a formula like Richter's since his diary for February 1855 records that he put 'sundry blue shadows' into the 'Chaucer' while it was in the 'gallery' of his dealer, 'Old' White.[123]

Pretty Baa-Lambs of 1851–59 (Figure 3.16) depicts intense blue–violet shadows on white surfaces, including the female figure's right shoulder, the baby's dress, and the fleeces of the sheep.[124] Brown worked on this painting 'in sunlight', as he recorded in his diary for August 1854.[125] And although he did rework it in 1853, and again in 1859, the shadows appear to have been completed during the initial campaign, which took place in the summer of 1851, since their distribution is identical in the small replica Brown made of the painting in 1852.[126] There is good reason, therefore, to believe

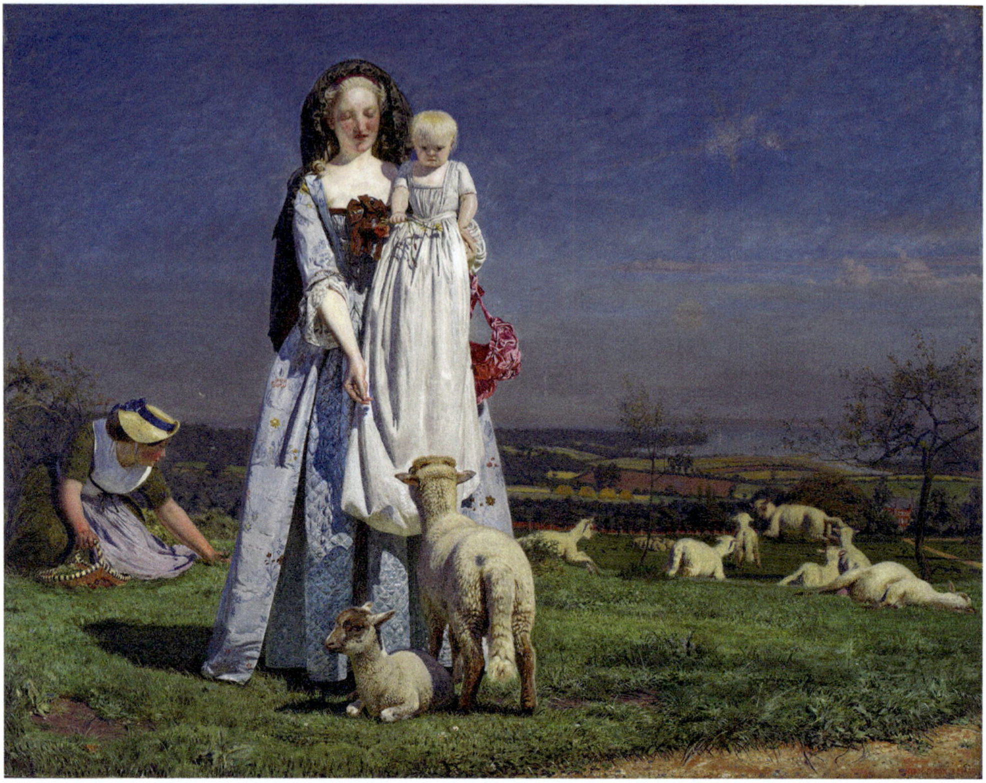

Figure 3.16 Ford Madox Brown, *Pretty Baa-Lambs*, 1851–59. Oil on canvas, 61 × 76.2 cm. Birmingham, Birmingham Museums and Art Gallery, acc. no. 1956P9.

Source: BMAG/Bridgeman Images.

that he could have seen coloured shadows of this kind. According to Marcel Minnaert, moreover, coloured shadows do appear on vertical surfaces under the conditions Brown represented since these are 'inclined' in such a way towards the high sun that it strikes them only 'grazingly', thereby reducing the contrast between light and dark that normally makes shadows look black.[127]

In 1855, the Pre-Raphaelite sculptor and poet of a 'scientific' bent, John Tupper, published a recollection of the blue shadows he had observed during daytime, while on a visit to the Dulwich Collection in his teens in 1837.[128] Citing his diary for the period, Tupper recalled how the 'shadows' that day 'were blue all along the Dulwich Road'![129] And he added: 'I had noticed these blue shadows, on a gravel path, some days before'.[130] The continuation of Tupper's account nonetheless reveals that these observations were influenced by a formula, since it recounts how his father, who was giving him 'instruction in optics' at the time, had told him that the 'shadows cast by sunlight were always, more or less, blue'.[131]

Hunt met Tupper around 1844, when he joined the Royal Academy Schools.[132] So it is not unlikely that the sculptor's ideas had some impact on the intense blue–violet colour of the shadows in *The Hireling Shepherd* of 1851 (Figure 3.17), which are

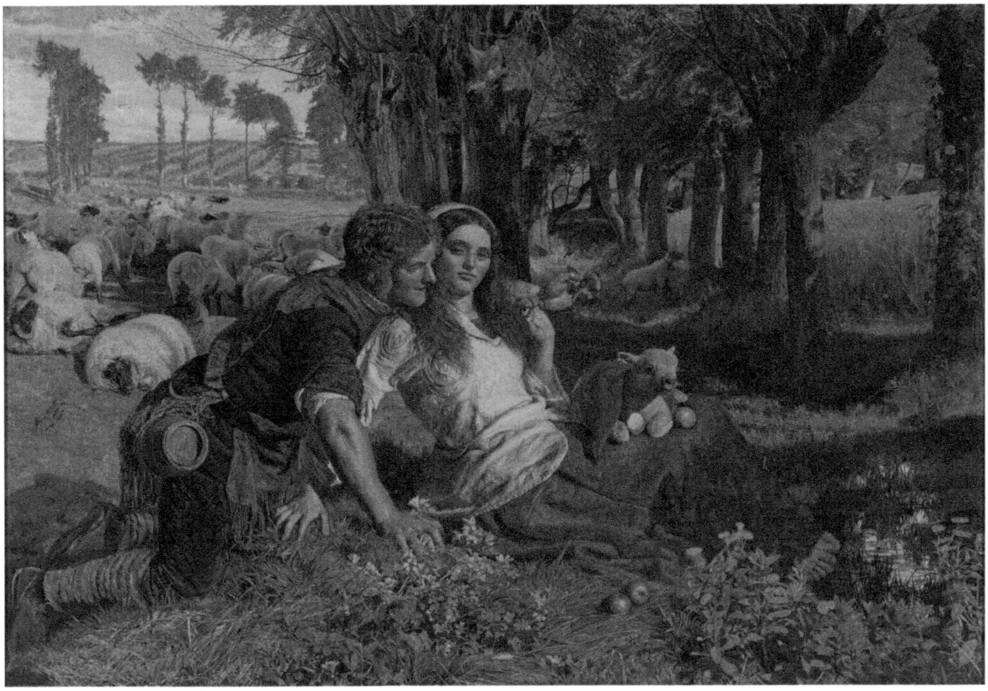

Figure 3.17 William Holman Hunt, *The Hireling Shepherd*, 1851. Oil on canvas, 110.4 × 144.5 cm. Manchester, Manchester Art Gallery, acc. no. 1896.29.

Source: Manchester Art Gallery/Bridgeman Images.

plainly visible in the woman's blouse and the man's hair and in the fleeces of the sheep and the grass in the meadow in the background. According to Hunt's own recollection, however, the painting fulfilled an ambition he had cherished since 1847 to make 'an out-of-door picture' by 'painting the whole out of doors, direct on the canvas itself', under the 'sunlight brightness of the day itself'.[133] Daniel Maclise was certainly convinced that Hunt had observed the coloured shadows he painted since he informed the painting's first owner, W.J. Broderip, in a letter of July 1852 that he had visited the site in Kent where it had been painted and had seen the 'very effect the artist had so well observed and portrayed', including 'blue shadows derived from the sky'.[134] Notwithstanding, it cannot be stated with certainty what time of day the work depicted because although the direction of its shadows indicates the light of early morning, Hunt worked on the painting *in situ* at Fitznells meadows near Ewell for several hours of the day.[135] He also worked on the landscape from June until October 1851.[136]

In the last analysis, it is probably true that most Pre-Raphaelite artists took it as an article of faith that shadows were blue. This may explain why they are visible in works by John Everett Millais, such as *The Woodman's Daughter* of 1850–51, which was not painted out of doors.[137] There was also a good deal of writing by this time in which the notion was a dogma. For example, the painter-turned-art theorist, Henry Twining, included a section on 'Shadows of Diverse Hues' in his *Philosophy of Art* of

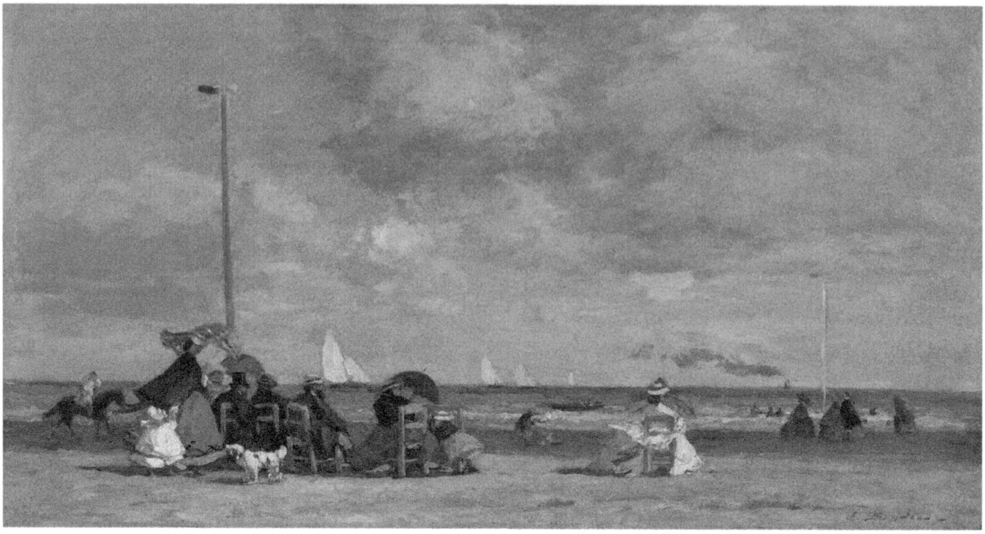

Figure 3.18 Eugène Boudin, *Beach at Trouville*, 1864–65. Oil on wood, 27 × 49.1 cm. Washington, National Gallery of Art, Ailsa Mellon Bruce Collection, acc. no. 1970.17.12.

Source: NGA Images.

1849, which stated that 'Even in the daytime, the shadows have mostly a bluish tinge, in open situations, because the principal light of reflection is that of the sky'.[138]

Blue shadows begin to appear in French paintings made during the daytime from the 1820s, in the work of Camille Corot, who may have learned about them from Valenciennes, and slightly later in the paintings of Johan Barthold Jongkind and Eugène Boudin.[139] In Boudin's *Beach at Trouville* of 1864–65 (Figure 3.18), for example, they are plainly visible on the white dress of the woman seated to the right of centre.

Boudin's importance to his informal 'pupil', Monet, is well-known. So it can be assumed that the older painter's example lay behind the blue shadows that began to appear in Monet's paintings of daylight scenes in the mid-1860s. In *Women in the Garden* (Figure 3.19) of 1866, for example, they manifest themselves on the white dresses worn by the female figures. It is nonetheless plain that Monet observed certain effects in this painting for himself, notably how the skirt of the woman to the left of the path was 'cut in two by the shadow and sunlight', as Zola remarked in his review of the 1868 Salon.[140] Monet was able to make such observations because he worked on the painting out of doors in the summer of 1866, recalling later that it marked the period when he 'threw' himself 'body and soul into the open air'.[141] And notoriously, he dug a 'trench' in his lawn and contrived an arrangement with pulleys – to Courbet's amusement – so that he could work on the upper portions of the painting on the spot.[142] (The fact that he reworked it over the following months does nothing to invalidate this claim.)[143]

Blue shadows also appeared in paintings which Monet's close companion, Renoir, made around the same time in bright sunshine. Notably, the shadows in the foreground of *The Pont des Arts* of 1867 (Figure 3.20), which are cast by unseen pedestrians overhead as they cross the Pont du Carrousel, are distinctly bluish in colour.[144]

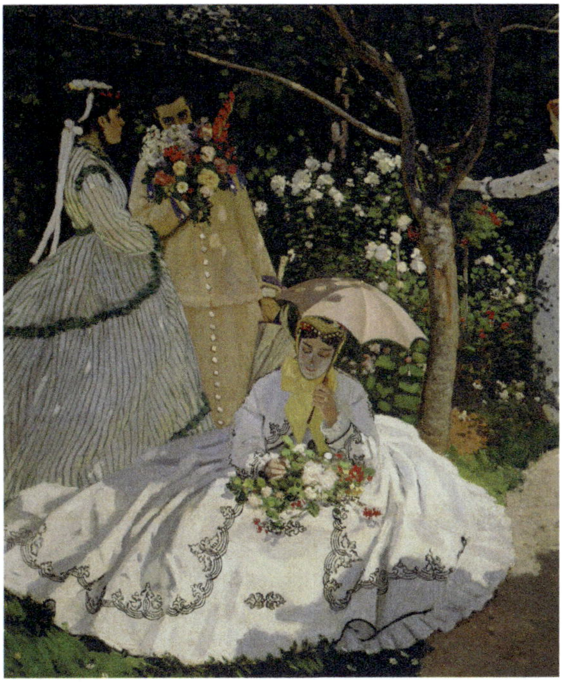

Figure 3.19 Claude Monet, *Women in the Garden*, 1866. Oil on canvas, 255 × 205 cm. Paris, Musée d'Orsay, inv. no. RF 2773, LUX 1360.

Source: Musée d'Orsay/Bridgeman Images.

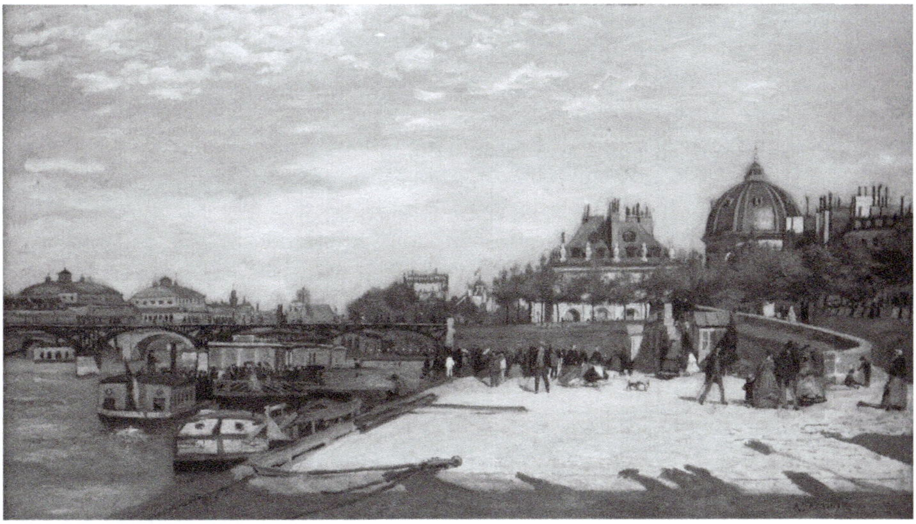

Figure 3.20 Pierre-Auguste Renoir, *The Pont des Arts*, 1867. Oil on canvas, 60.9 × 100.3 cm. Pasadena, The Norton Simon Foundation, acc. no. F.1968.13.P.

Source: Author.

It was only many years later, however, that Renoir gave a clue to his interest in such effects, when in 1910 he told a young painter who used white to represent snow:

> White does not exist in nature. You admit that you have a sky above that snow. . . . Your sky is blue. That blue must show up in the snow. . . . Shadows are not black; no shadow is black. It always has a colour.[145]

This dispute conveniently sums up how coloured shadows only appeared to some observers during the daytime. The reason for this, it would seem, is that we do not all see blue shadows equally well. For one thing, physical factors, such as variations in the number of short-wavelength-sensitive cone cells in the retina, can dispose some people to be more sensitive to them than others.[146] Higher-level perceptual mechanisms also give different individuals a unique sense of what is white, as opposed to blue or yellow, as a result of which they will register the colour of a shadow differently.[147] More significantly, perhaps, although we have evolved to eliminate variations in the colour of the illumination and preserve object colour, we factor out blue light much more readily than yellow light.[148] So, while we tend to regard yellow tints on a surface as local colour and tend to see them as such, we are prone to discounting blue tints as accidental effects of light and to failing to register them.[149] All this implies that we must make a special effort to see the blue colour of the shadows that are potentially visible under sunlight. If so, then what artists and scientists believed had a crucial effect on what they expected to see, on what they looked for, and what they actually saw.

A Perfect Mirror

In some situations, when skylight does not produce a blue shadow on a matt surface, it will do so on a shiny surface, as this will reflect much more blue light. Although it is debatable whether phenomena of this kind really are shadows, it is clear that some artists thought they were, not least because they could be as seen as reflections.

Brown clearly believed this was the case with the shiny surfaces lying in shadow in the foreground of *An English Autumn Afternoon* of 1852–53 (Figure 3.21). This painting, he explained in *The Exhibition of Work*, represents effects he observed at '3 P.M., when in late October the shadows already lie long', initially while painting from the window of his Highgate studio.[150] It shows in particular how the 'upper portion of the sky would be blue as seen reflected in the youth's hat'.[151] The same reasoning would also explain why the top surfaces of his companion's satin dress are spattered with blue and perhaps why her parasol appears blue. Importantly, however, the matt surfaces next to these items, notably the white lace trimmings on the woman's dress white and the seat of the nearby bench exhibit no blueness at all, even though they are top-lit by the same blue sky.

Delacroix made a similar observation to Brown in his journal for August 1854. Describing the shadows he saw on the shiny surfaces of the waves at Dieppe during his early morning walk, he wrote:

> I studied the sea for a long time. The sun being behind me, the faces of the waves that stood up in front of me were yellow, and those facing the other way reflected the sky . . . in the background, where the sea was blue and green, the shadows seemed violet.[152]

Delacroix had in fact already recorded effects of this kind in *The Sea Viewed from the Heights of Dieppe* of 1852 (Figure 3.22), a sunset scene painted from memory in

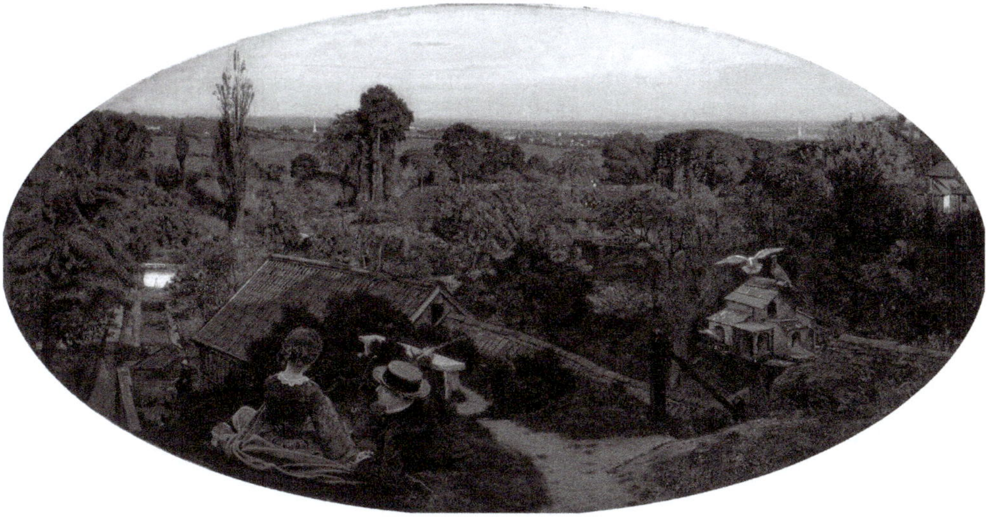

Figure 3.21 Ford Madox Brown, *An English Autumn Afternoon*, 1852–53. Oil on canvas, 71.7 × 134.6 cm. Birmingham, Birmingham Museums and Art Gallery, acc. no. 1916P25.

Source: BMAG/Bridgeman Images.

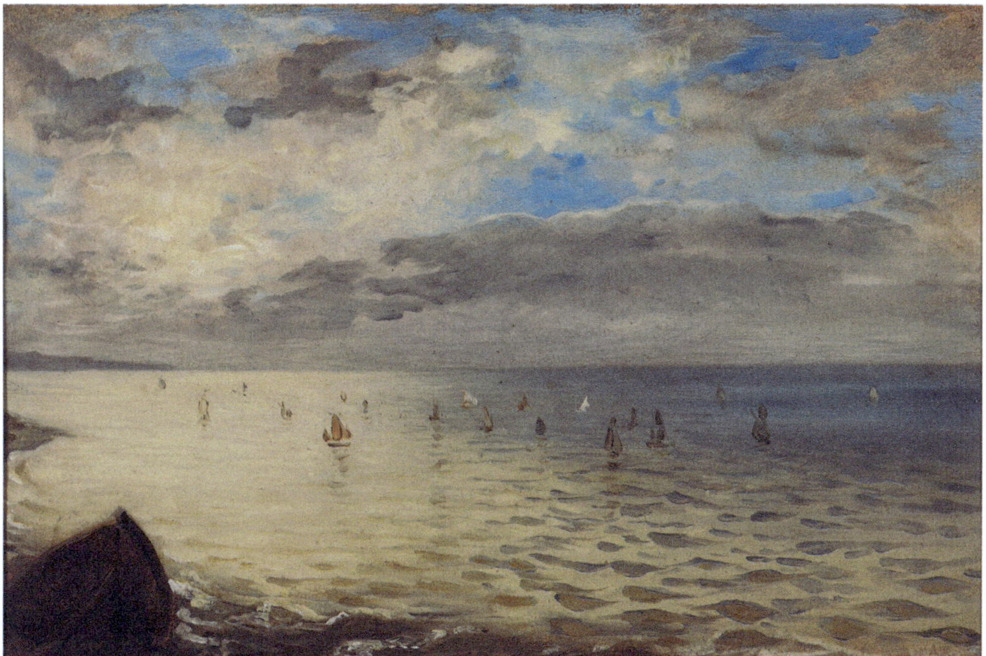

Figure 3.22 Eugène Delacroix, *The Sea Viewed from the Heights of Dieppe*, 1852. Oil on canvas, 36 × 52 cm. Paris, Musée du Louvre, inv. no. RF1979–46.

Source: Musée du Louvre/Bridgeman Images.

September.[153] Here, that is, the opposing faces of the waves are painted in strokes of warm paint when they reflect the sun and are painted in strokes of cool paint when they reflect the sky.

No painting of Samuel Palmer's survives which indicates his belief that the shadows on shiny surfaces reflected the sky, although this is clearly implicit in a letter of 1856, where he wrote:

> Beware of the notion that shadows cannot be cast upon water. It is, I believe, true of a perfect mirror, but when is water such? I lately saw the shadow of a pier cast upon the sea, and its colour thereby totally altered by losing the warm sunlight, and reflecting to the eye only the blue of the sky above.

Anomalies

Although physics seemed to explain most shadow colour phenomena plausibly enough, it ran into the same trouble as its Aristotelian predecessor when it came to explaining how a shadow could appear coloured even when illuminated only by a colourless light source.

One of the most interesting, although finally abortive, attempts to explain this phenomenon came in Karl Scherffer's book, *Abhandlung von den zufälligen Farben* of 1765, which was nominally devoted to 'accidental colours' or coloured after-images of the kind described by Buffon.[154] Here, though, Scherffer made several observations about coloured shadows. In one of these, he described how a 'body' placed on 'white paper' illuminated by daylight and a 'lamp or a lighted candle' created two shadows, 'one cast by the daylight' which was 'yellow' and the other cast 'by the lamp and fully lit by daylight' which was 'quite blue'.[155] His explanation for both phenomena was straightforwardly physical. Thus, the blue shadow is blue because the daylight illuminating it is always blue, that is not only when the 'sky is serene' but also when it is 'covered in clouds'.[156] This argument is simply mistaken, of course. Worse, it shows that Scherffer failed to draw the lessons from other experiments of his that could have explained why shadows appear blue even when lit by neutral daylight.

Among these was one which he conducted to 'find out' whether the after-image produced by a 'truly blue' shadow was 'yellow', as it was with 'bodies that have this colour'.[157] To this end, Scherffer fixed his attention 'for a long time' on a blue shadow cast by a 'lit lamp' and then moved his eyes to find that the 'white paper' did in fact display the 'entire shape' of the shadow in a 'golden yellow colour'.[158] Similarly, he observed that a yellow shadow produced a 'violet-blue' after-image.[159] Scherffer thus demonstrated that coloured shadows do indeed engender successive contrast effects, just as normally coloured surfaces do.[160] However, rather than seeing this phenomenon for what it was, namely an effect of adaptation, Scherffer took it to confirm his belief that coloured shadows were physically blue.

Scherffer's more general theory of successive contrast was more successful. It is distinguished by the fact that it was the first to trace the cause of after-images to 'fatigue' in the retina and to argue that the colour circle (suitably revised) could predict the hues of 'accidental colours'.[161] It also investigated the effect of the light emitted by the background colour on the eye's sensitivity to after-images.[162] But despite having demonstrated the operations of what we now call successive contrast in such a rich fashion, Scherffer did not think to ask whether this process played any role in generating coloured shadows – although some later investigators suggested he did.

An even more remarkable aspect of Scherffer's account of accidental colour is that it elaborated several key elements of a theory of constancy, which anticipated Monge's and could have been extrapolated to explain coloured shadows. A foundation of this was the observation that the 'light of a candle' was in fact, or objectively, 'yellow' rather than colourless.[163] Scherffer then explained how we

> nevertheless regard this [yellow] light as white only for the reason that we also attribute this colour to bodies which, being compared with others which we see at the same time, return, not only a greater number of rays, but also return them more evenly mixed.[164]

Although cryptic, this sentence makes three important points. The first reiterates a principle Scherffer had elaborated at the outset of his book that 'on everything relative to our judgment on colours', the 'inductions we draw' are 'mainly based on the comparison of the colour that we are considering, & of the other colours that, at the same time, strike our eyes'.[165] (This idea, that objects have the colours they do because we compare the light they reflect, was later a central plank of Monge's theory of constancy.) Scherffer's second point was that yellow objects look white under a yellow light because they reflect more rays of light than objects of other colours, implicitly just as white objects do in most illuminations. (And again, Monge made exactly the same point later.) Scherffer's third point was that yellow objects look white under yellow light because in this situation they reflect all the different kinds of light rays in equal proportions.[166]

Scherffer rather confused matters in the continuation of this passage. It is nonetheless clear enough he meant to say that we habitually adjust the colour of the illumination to white when he described how: '[I]n the evening, when an apartment is lit by candles, we do not see any change in the colours of the things there', although – objectively – 'they turn yellow'.[167] The same conclusion is indicated by his observation that 'By day, & mainly after twilight, a blue colour spreads over all bodies; but we seldom perceive this, for lack of a body which, being exempt, can serve as an object of comparison.'[168]

Although Scherffer did not use this theory of constancy to explain coloured shadows, how he could have done so is demonstrated by Mollon's extrapolation of Monge's argument. That is, in adjusting the dominant illumination to white, or away from yellow, we adjust shadow colour accordingly, or towards blue. And although there is no mention of Scherffer in Monge's writings, it is surely no coincidence that *Abhandlung von den zufälligen Farben* was translated in its entirely into French in the *Journal de Physique* in 1785.[169]

Beguelin was also intrigued by the fact that a combination of moonlight and candlelight could generate coloured shadows. His interest in this phenomenon was sparked by a paper 'Observations sur des couleurs engrendrées par le frottement des surfaces planes et transparentes', given by the Abbé Guillaume Mazéas in 1752, which set out an experiment in which 'Moonlight and candlelight were both made to strike an opaque body'.[170] The result, Mazéas observed, was that the 'shadow the opaque body created where it intercepted the moonlight was red, and the shadow the same body created where it intercepting the candlelight appeared blue'.[171] Mazéas attempted to explain these phenomena with the Aristotelian argument that that these 'colours arise simply from the enfeeblement of light'.[172] Beguelin, however, stated that this account was neither 'clear, nor satisfactory'.[173] He also dismissed what he took to be Mazéas's argument that the 'coloured shadows' in this experiment were produced by the 'attraction' the opaque body casting them exerted over the light rays falling in the shadow and more specifically by the 'distraction' or 'reflection' of these rays this entailed.[174] Unfortunately, Beguelin's own explanation was no more coherent, since it rested on the mere assertion that the 'blue of sky' somehow found its way into, and mixed with, the moonlight illuminating the shadow.[175]

The rarely seen but striking green shadows that are sometimes visible outdoors at dawn and dusk presented an even greater anomaly to physical theory. The best Leonardo could do was to argue in Chapter 158 of the *Trattato*, with reference to a diagram (Figure 3.23), that 'an opaque body' can exhibit 'greenish shadows and reddish lights' when

> light . . . from the east . . . illuminates that object with the colour of its glow while, in the west, an object of a colour different from the first is illuminated by the same light. Its reflected lights will rebound eastward, and its rays will strike the parts of the first object facing it, whereupon those rays will be truncated as their colour and brightness affixes to the object.[176]

For all its ingenuity, this explanation is unserviceable since no object can reflect green light in any quantity when the sun is low, as is implicit in this explanation, since it is almost devoid of medium-wavelength light at such times.

Two centuries later, Buffon described 'green shadows' cast by trees on a 'white wall' in July and the 'tender green' shadows, as if 'newly painted in verdigris' cast by a trellis on the same wall.[177] Significantly, he noted that they appeared when he was 'in an elevated place', and the 'strongly red' sun was 'setting' in a gorge so that it was 'considerably below the horizon'.[178] Indeed, Buffon seems to have realised that the sun's

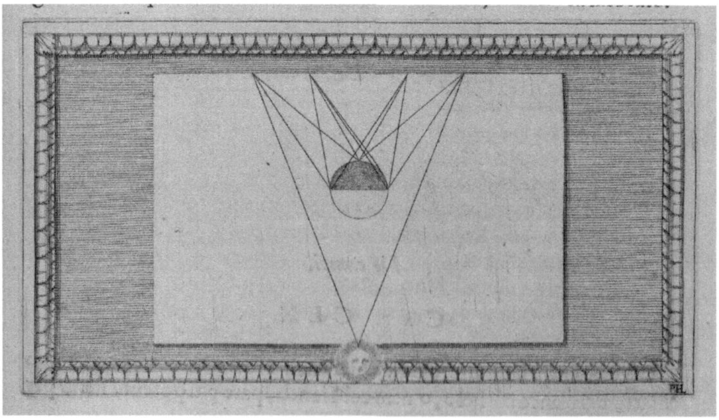

Figure 3.23 Green shadows. Illustration to Chapter 158 of Leonardo da Vinci, *Trattato della pittura* (Paris: Jacques Langlois, 1651b). Engraving by René Lochon.

Source: Author.

position mattered to the visibility of green shadows since he noted how the shadows appeared 'blue, or rather the colour of indigo most lively' at dawn the next day when the sun was already 'elevated above the horizon'.[179] And, subsequently, he observed how they were a 'beautiful azure blue' when the sun setting disappeared behind a rock before it could descend beneath the horizon'.[180] Buffon noted, in addition, that the shadows were green when there was also a 'transparent curtain of reddish yellow vapours' in the atmosphere, but blue when there were no 'yellow vapours'. The suggestion implicit on these observations, that green shadows are produced by yellowish light falling into blue shadows, is mistaken, however.[181] This is because the additive mixture of yellow light does not mix with blue light to make green, unlike the subtractive mixture of yellow and blue paint, but creates white.[182] And orangey light mixed with blue makes a faintly pinkish white.

These facts did not prevent Beguelin from reaching similar conclusions in his gloss on Buffon's observations. Working from the premises that 'shadows' are 'naturally blue' when the sun is low and that 'green is only a compound of blue and yellow', he argued that a green shadow is produced when 'something yellow is mixed with [a] blue shadow'.[183] And, predictably, he claimed that a green shadow occurred when the 'blue rays' a shadow received from the 'atmosphere' combined with the 'light' from a 'transparent curtain of reddish-yellow vapours', much as Buffon did.[184] He also suggested that one could appear if 'yellow rays' of light fell into the 'shaded section' of a white wall when this was illuminated by the 'colour of the sky'.[185]

Buffon's argument was tenacious. Like Beguelin, H.F.T. claimed that the 'green' colour of the shadows his predecessor had seen was a 'composite' of 'blue' skylight with the 'red mixed with yellow' light emanating from 'vapours' in the sky.[186] What is more, a passage in Valenciennes's *Elémens de perspective pratique* indicates he knew Buffon's argument, or one of its derivatives, since this states that 'physicists have tried to explain the colour' of a green shadow by 'attributing it' to 'red vapours etc.'.[187]

By the early nineteenth century, the primary coloured lights had in fact been correctly identified more than once as red, blue (or violet), and green (rather than yellow). But it was not until much later, in the 1860s, with the publication of the work of James Clark Maxwell and Hermann von Helmholtz, that scientists and a few artists understood the distinction between the additive and subtractive primaries.[188] As a result, the idea that yellow and blue, whether in the form of light or pigment, made green remained conventional wisdom in both camps for some considerable time. It is no surprise, then, that Beguelin's argument was echoed by Xavier de Burtin, albeit unclearly, in his *Traite théorique et pratique des connoissances qui sont nécessaires à tout amateur de tableaux* of 1808. This maintained that because the 'colour of shadows' depends on the 'illumination' and 'consequently' the 'atmosphere', 'red' and 'yellow' light from these can modify the 'blue' that 'is always the base of a cast shadow' to the extent of rendering it 'more or less greenish or *violet*'.[189]

Beguelin gave a second explanation of green shadows – also based on subtractive mixture – when he suggested they could occur if a 'blue shadow' fell on a surface of a 'yellowish' or 'yellow' colour.[190] H.F.T. later reiterated this argument, arguing that a 'green' shadow is produced whenever a 'blue' shadow falls on a 'yellow' surface or a 'yellow shadow' falls on a 'blue' surface.[191] Similarly, Valenciennes contended in the *Elémens de perspective pratique* that 'If the colour of a body is yellow, then the shadow will be more or less green, as a result of how the transparent blue shadow, mixing with the yellow, will inevitably form green'.[192] Valenciennes mentioned that he gave this 'explanation' as an alternative to those proposed by 'physicists', using the 'knowledge provided by the Art of Painting about the application of transparent colours on to those that are not, and about the hues that result'.[193] The problem here, however, is that a yellow surface absorbs most of the 'blue' (short-wavelength) light that falls on it, just as a blue surface absorbs most of the 'yellow' (medium- plus long-wavelength) light that it receives, with the result that neither appears green.

Beguelin's argument was nevertheless taken up again by Goethe in the *Farbenlehre*, where he stated that if 'we produce [a] blue shadow' on 'light yellow paper, the surface will appear green'.[194] This passage was subsequently included in the English edition of 1840, which was extracted almost exclusively taken from the first, 'didactic' volume of the original.[195] The Pre-Raphaelites were probably aware of this text, since they included Goethe among the 'immortals' in the manifesto they issued in late 1848.[196] It is possible, then, that Goethe's observation was a factor in Hunt's decision to colour the long, evening shadows cast by the corn stooks green in his small painting, *Cornfield at Ewell* of 1849 (Figure 3.24). This work was executed in large part out of doors, from observation.[197] The fact, however, that Hunt added different quantities of green paint to these shadows and scraped through the areas between them to reveal the corn underneath could suggest he believed they were created by blue light from sky falling on its yellow surface.

Beguelin tentatively offered a third explanation of the green shadows seen by Buffon: that they were caused by light reflected off the white wall onto the 'neighbouring greenery' returning 'coloured with green on to the wall'.[198] He nevertheless admitted

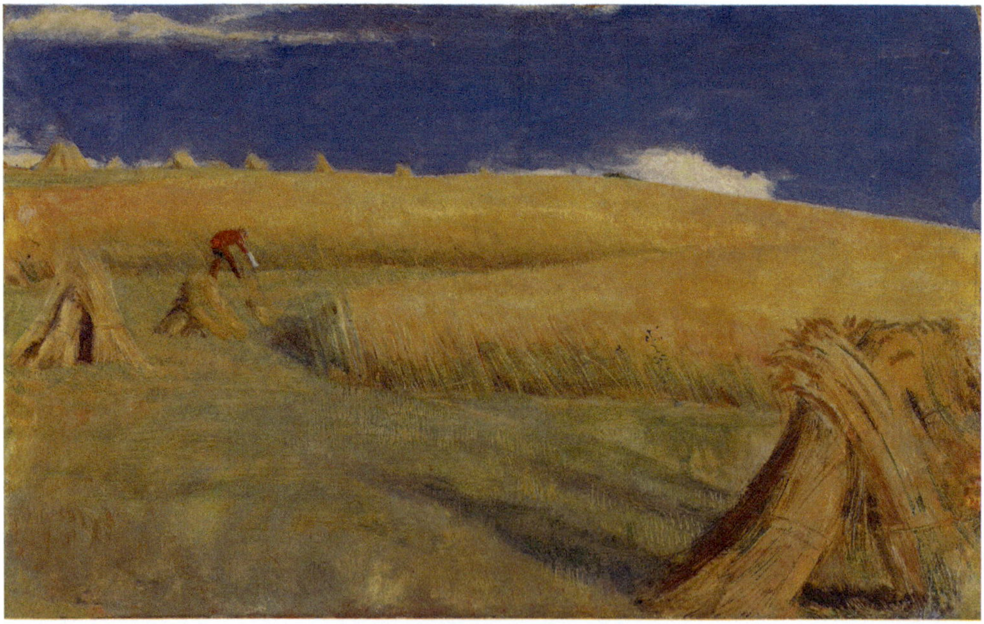

Figure 3.24 William Holman Hunt, *Cornfield at Ewell*, 1849. Oil on millboard, 20.2 × 31.8 cm. London, Tate Britain, ref. no. T05468.

Source: © Tate.

that he 'never saw this green reflection', although he 'expected' to do so when observing some 'trees next to a white wall opposite the setting sun.'[199]

Valenciennes later proposed a slightly different explanation of green shadows in the *Elémens de perspective pratique*, which he based on the different idea that foliage filters the light falling on it. Thus, he argued that 'Transparent leaves which intercept the sunlight will produce a greenish shadow' with a slight 'golden' cast.[200] This belief (or Beguelin's) may explain why some of the shadows near the trees in the architectural background of *The Ancient City of Agrigentum* have a muted green tinge – even though the light that passes through (or reflects off) foliage is too weak to produce this result.[201]

While the foregoing accounts do at least have a certain plausibility, this cannot be said of the explanation of green shadows offered by Adam Walker in his *A System of Familiar Philosophy* of 1799. (Although a friend of Priestley, Darwin, and other Lunar Men, Walker's understanding of physics was self-taught, and his book was a potboiler intended to promote his career as an itinerant lecturer.)[202] Here, Walker began conventionally enough by explaining how the blue 'colour of shadows on a white ground' is caused by the action of a 'blue sky'.[203] He then launched into his account of the green shadows he produced in an experiment involving a red filter. Referring to a diagram (Figure 3.25), this stated:

> A very surprising effect of shadow is produced by placing a red flat piece of glass, *a* . . . before the candle *d*, and holding a pencil, *c*, between it and the sheet of white paper *x x*. If another candle, *g*, be placed a few inches from the first, two shadows of the pencil will be produced on the paper, *x x*; that from *d*, through the red glass *a*, viz. *e e*, will be a fine green; and that from the candle *g*, viz. *n x*, will be a deep red.[204]

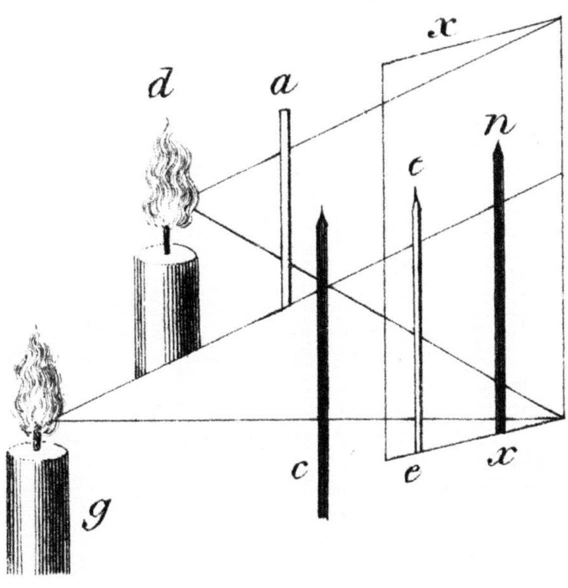

Figure 3.25 Coloured shadows. Fig. 61, Plate VII from Adam Walker, *A System of Familiar Philosophy* (London: Kearsley, 1799), p. 102.

Source: Author.

For Walker, the problem at issue was that the 'interception of red light should produce a green colour' in the shadow illuminated only by unfiltered and implicitly colourless candlelight. Although Rumford has already explained shadow colours of this kind as a function of contrast, Walker made no mention of his ideas or experiments.[205] Instead, he argued that the shadow he had seen was not really green at all but had a blue colour which happened to 'appear green', rather – he added unhelpfully – as 'strong white light' does when 'looked at through blue glass'.[206] Walker then continued down the same path by suggesting that, just 'as shadows on a white wall are of a dubious blue', might not the 'near approach of blue and green to one another, make one, sometimes, to be mistaken for the other'?[207]

This explanation is a good example of how an inadequate theory can blind its adherents to the phenomenology of coloured shadows. The point is that the green colour shadows under the circumstances Walker described is vivid and unmistakeable, as in the photograph of Miranda in the Introduction, and not in the least 'dubious'. Walker's confusion may nevertheless explain why so few artists painted green shadows, if – like him – they could not see them as the upshot of a process which they could readily grasp.

Notes

1. Nollet 1743–64, 5: 514. See also H.F.T. 1782, 157–159, which cites both passages.
2. See Major 1761 (citing the supplement to the *Ephémérides des curieux de la nature* of 1 December 1672). Major's ideas were also paraphrased in Chaptal 1790, 3: 145.
3. Major 1761. See also Lairesse 1738, 1: 210 for a description of how he had identified coloured shadows by making a 'small Hole in the Window of a darkened Room' which projected them on to a 'white Wall'.
4. da Vinci 1651 (French edition), 7.
5. Ozanam 1788, 2, 283.
6. Ibid.
7. Bouguer 1760, 70. See Baxandall 1995, 108, on this theory.
8. Bouguer 1760, 366.
9. Ibid., 367–368. Cited in Baxandall 1995, 113.
10. Newton 1704, 60.
11. Melvill 1756, 2: 12–90, esp. 74.
12. Valenciennes 1799, 285. Cited in Gallo 2017, 183. See Valenciennes 1799, 287–288.
13. Ibid., 254.
14. Beguelin 1769, 32.
15. See ibid., 35–37.
16. Hagedorn 1775, 2: 190.
17. For an overview of Delacroix's interest in coloured shadows, see Schöller 2017, 55–65.
18. Delacroix 1893, 1: 143. For a searching analysis of Delacroix's ideas about half-tones and their relation to his ideas about coloured shadows, see Roque 2009, 251–253. See also Johnson 1981–89, 3: 98.
19. Delacroix 1893, 1: 158.
20. Delacroix 1893, 3: 201. On Delacroix's ideas about the role of reflections in colouring shadows, see Schöller 2017, 52.
21. See Delacroix 1893, 2: 302, for a journal entry of April 1860 stating how a 'commentary on Leonardo's treatise on painting would be an unusual work'.
22. Duranty 1876, 22.
23. Ibid.
24. Rivière and Schnerb 1978, 88. See also Bernard 1904, 24, for Cézanne's statement that 'Shadow is a colour like light, only less brilliant'. For a description of Monet's practice with respect to coloured shadows more generally, see Schöller 2017, 89–109.

25. See Ratcliffe 1960, 330–331, on the inscription mentioning Régnier's book on the inside cover of a Cézanne sketchbooks (Chappuis II) datable to 1877–85. This cover was lot 562 in the Christies sale, Impressionist and Modern Works on Paper, of 9 February 2006. See also Beucken 1955, 304, which states that Cézanne's copy of Régnier was on the shelves of his grandson, Jean-Pierre Cézanne, at the time of the book's publication.
26. Régnier 1865, 69.
27. See Rewald 1986, 96; and Renoir 1962, 111.
28. Pissarro 1980–91, 1: 260 [no. 198]. Cited in Watson 2004, 24. The copy belonged to Armand Guillaumin. See also Pissarro 1980–91, 5: 337 for a letter of May 1903 from Pissarro on how Turner and Constable 'had not understood "the analysis of shadow", which in Turner's painting always tends towards an effect, a void'.
29. On the blue shadows in Impressionist snow scenes, see Duret 1878, 16. Cited in Roque 1996, 35 and Roque 2009, 305. For an alternative description of the shadows in this painting, see Schöller 2017, 96–97.
30. On these manuscripts, see Farago et al. 2018, 1: 15–20 and 322–328. The shadows in Mauperché's painting are rendered in grey paint but acquire this colour by contrast with the warm tones around them.
31. Valenciennes 1799, 249–250. Cited in Gallo 2017, 182. For a discussion of Valenciennes's insistence on the scientific observation of 'causes', see Strick 1996, 81.
32. Valenciennes 1799, 417–418; cited in Strick 1996, 86.
33. Valenciennes 1799, 288, 407 and 292. For a discussion, see Strick 1996, 85.
34. See Valenciennes 1799, 577–579 for the artist's description of the site.
35. Valenciennes 1799, 381–385 and 479–482. On Valenciennes's trip to Agrigento, see Lui 2009, 204.
36. Ibid. 285–286. Cited in Cavina 2001, xli and Gallo 2017, 183. See also Valenciennes 179, 400, for a mention of Leonardo's writings on anatomy.
37. See, for example, Scherffer 1765, 5–7 and 39–40; Gehler 1790, 2: 824–826; and Voigt 1796, 271–275.
38. Runge 1840, 1: 82.
39. Siegel 1978, 19.
40. For an overview, see Conisbee et al. 2003, 70–71 and Hedin 2015, 171 n. 4. See also Monrad 2015, 20, for the suggestion that Eckersberg derived his interest in shadows more generally from Lairesse and Christensen 2013, 262, 363, and 578–583 on Eckersberg's close relationship with the scientist, Hans Christian Ørsted, from 1812. On how Eckersberg arranged for students of the Royal Academy of Fine Arts to attend the lectures Ørsted gave in 1824–25, see Christensen 2013, 581; Monrad 2015, 35; and Hedin 2015, 163.
41. See ibid., which states that Thorvaldesen, with whom Eckersberg stayed while in Rome, also owned a copy of this book. See also Monrad 2003, 16 and Monrad 2015, 24.
42. See ibid., 18 and 52 n. 34; Conisbee 2003, 40–41 and Hedin 2015, 154. On Eckersberg's interest in Valenciennes's student, Chauvin, whose studio he visited while living in Rome, see Monrad 2003, 16; and Monrad 2015, 24–25 and 33.
43. Ibid. 2015, 49.
44. Ibid., 27–28.
45. Cited in Conisbee et al. 2003, 74–75.
46. Although the blue colour of these shadows is enhanced warm tones depicting highlights created by the setting sun, Wright used a distinctly bluish colour to depict them in the first place.
47. Priestley 1772, 2: 438–445, which also mentions Mazéas's and Bouguer's arguments.
48. See Wright 1988, 26 and 29.
49. See Robinson 1962, 156; Schofield 1966, 144–146, 149, and 153–154.
50. See Schofield 1963, 130–131.
51. See Nicolson 1968, 1: 18. See also Fraser 1990 on Wright's association with the Lunar Society, and his friendship with Darwin. On Darwin (who left Birmingham shortly after Priestley arrived) and the Lunar Society, see Robinson 1962, 156 and 162 and Schofield 1966, 144–149, 154, 157, and 159.

52. On Darwin's move to Derby and his involvement with the Derby Philosophical Society, see Uglow 2002, 351 and 377.
53. See Darwin 1791, 1: 8 (of the 'additional notes' following 214).
54. See Schofield 1966, 147–148, and 154 and Uglow 2002, 320–321, 377, and 481.
55. Edgeworth 1800, 151–152.
56. Edgeworth 1825, 4: 175–177; citing Priestley 1772, 2: 438; citing Leclerc 1746, 157–158. Coloured shadows are also mentioned in Edgeworth 1798, 1: 57 and Edgeworth 1806, 1: 152–153.
57. Richter 1817, 3–6, 15, 61. See also ibid. 7, on the 'mild reflection of the sky'. The book is cast in the form of a dialogue between the author and the ghosts of the painters, Cuyp, Rembrandt, Rubens, Teniers, and Van Dyck, which takes place in the 'Exhibition of Dutch and Flemish Pictures at the British Institution' of 1815. See Anon 1815. See also Roget 1891, 388, which cites a certain J. W. Papworth, on how the 'bluish reflected hue' in Richter's draperies was intended 'to reflect in a strong degree the over-hanging blue of the atmosphere'. On Richter's book, see Gage 1969, 19–20; Staley 1973, 29; and Kemp 1990, 304.
58. Richter 1817, 6.
59. See Anon 1812. See also Story 1894, 32, for a description of a different scenario in which the work 'was painted on the top of the house in which he then lived in a blaze of sunlight', causing 'Mr. Chamberlayne' for whom the 'picture was executed' to remark 'on the exposure of his models to such fierce sunshine'. The earliest reproduction of the painting is a mezzotint of 1816 by John Young. Roget 1891, 1: 386 states that a 'repetition' of the painting was exhibited at the Society in 1816, the original having been shown in 1812 'at the rooms of the Associated Painters, in Old Bond Street' and that 'One or the other' was 'afterwards placed over the altar of Greenwich new church'. See also O'Keefe 2009, 103.
60. Richter 1817, 8 and 10.
61. Roget 1891, 1: 386 (drawing on notes compiled by Joseph John Jenkins, sometime secretary of the Old Water-Colour Society of which Richter was a member). See also Story 1894, 27 for the claim that 'Richter was one of the first "to go to nature"'.
62. Richter 1817, 66 (part of a discussion of 'Kantesian theory'). See also Roget 1891, 385 on Richter's Kantianism. For an account of Richter's article on metaphysics in the *Encyclopaedia Londonensis* of 1814, see Sartain 1899, 124.
63. Richter 1817, 23.
64. Ibid., 15
65. See Gilchrist 1863, 1: 295 and Roget 1891, 1: 388 on how both artists visited John Linnell's cottage at Hampstead in the mid-1820s.
66. See Worrall 1975, 405.
67. Reynolds 1798, 1: 275. Cited in Blake 1988, 662. Blake wrote on the title page: 'Hire Idiots to Paint with cold light & hot shade'. See also Saint-Morien 1788, 150, for the argument that 'violet' was 'the colour that most closely approaches the dominant tone of shadows'.
68. *Houding*, translated as 'harmony', is discussed in Lairesse 1738, 1: 174–177. On *houding*, see Taylor 1992, 215–220 and De Vries 2004, 86. On *cangiante* colour, see Cennini 1821, 73–74 and Bomford 1995, 11. See also Gilchrist 1863, 369, on how Linnell gave Blake the 'first copy of Cennino Cennini's book seen in England'.
69. See Willson 1842, 26, 37, 54–55, for citations from Rigaud's translation of the *Tratatto*.
70. Ibid., 53–54 and 56.
71. Ibid., 54; citing da Vinci 1835, 14.
72. Willson 1842, 54.
73. Willson 1842, 53. See also ibid., 57–58 on the importance of adapting shadow colour to the unity and 'harmony' of the picture.
74. Stephens 1860, 19–20. Cited in Staley 1973, 24.
75. Hunt 1905, 2: 471. Cited in Kemp 1990, 304.
76. See Hunt 1886, 483 and Hunt 1905, 1: 173 and 183–184. See also Staley 1973, 11 and Bronkhurst 2006, 1: 134.
77. Beguelin 1769, 33 and 36–37.
78. H.F.T. 1782, 53 and 56. See also Leclerc 1774, 339 for a letter from the Abbé Millot, which describes the 'red' shadows (among others) which form around a blue shadow.
79. Carus was trained as a physiologist. See Carus 2002, 1–3.

80. Carus 2002, 145.
81. Willson 1842, 58.
82. Ibid.
83. I am grateful to Caroline Arscott for this observation.
84. Willson 1842, 58.
85. Régnier 1865, 44.
86. Ibid., 47. I am grateful to Teresa Bridgeman and Alexandra Parigoris for this translation.
87. Ibid.
88. Wildenstein 1974–91, 3: 285. Cited in House 1986, 198.
89. Minnaert 1954, 245.
90. Brücke 1866, 200.
91. Ibid.
92. Farago et al. 2018, 2: 680–681. See also da Vinci 1540, f63v; and da Vinci 1651, 27.
93. da Vinci 1540, f229v. Translated in da Vinci 1956, 1: 264.
94. Farago et al. 2018, 2: 713. See also da Vinci 1540, f74; and da Vinci 1651, 40.
95. Farago 2108, 2: 716. See also da Vinci 1540, f75; and da Vinci 1651, 41.
96. Lairesse 1738, 1: 194.
97. Ibid., 1: 210. See also ibid., 1: 11 and 191.
98. Ibid., 1: 210.
99. Leclerc 174, 336.
100. Beguelin 1769, 31–32.
101. Ibid.
102. Churma 1994, 4721. See also Hardin 1988, 25 on the wide range of light–dark contrast visible in full sunshine.
103. Cited in Leclerc 1774, 337.
104. Ibid., 338.
105. H.F.T. 1782, 20–21. For H.F.T.'s statements on the principle concerned, see ibid., 7, 19, 21, 26, 28, 32, 39, 45, 48, 53, 67, 75, 77, 109, and 124.
106. Ibid., 18–19 and 21.
107. Ibid., 22.
108. Ibid., 22–23. See also ibid., 39–40, for a description of a similar case.
109. Valenciennes 1799, 285.
110. Ibid., 289.
111. Saussure 1779–96, 4: 296.
112. See Hansen 2013, 111 and Anon 1914. Méchel made these etchings from drawings by Marquardt Wocher, which were themselves elaborated from sketches Henri Levêque made in 1788, over a period of several weeks while Saussure's party was camped out on a pass between Chamonix and Courmayeur.
113. Melvill 1756, 75.
114. Darwin 1791, 1: 8 (of the 'additional notes' following 214).
115. Brougham 1796, 255.
116. Ibid.
117. Richter 1817, 11.
118. See Minnaert 1954, 245.
119. Palmer 1974, 2: 961. Cited in Roget 1891, 1: 387–388. See Palmer 1974, 1: 533, for a letter of June 1857, in which the artist told Linnell of his familiarity with the house in Newman street where Richter had lived in the 1810s and 1820s. See also Story 1894, 32; and Sartain 189, 112.
120. See Kemp 304, which suggests a link between the 'striking blue reflections' in the painting and Richter's theory. Brown recorded in his diary that he conceived this painting (which went through several versions) in 1845, began it in 1847, and subsequently painted it during the summer of 1849, before he exhibited it in 1851. On the key stages of the campaign, see the entries for 4 September, 1847, 4 May 1848, and 16 August 1854 in Brown 1981, 1–4, 40, and 73. See also Bennett 2010, 54–59.
121. Brown 1865, 3. See also Staley 31 and Treuherz et al. 2011, 132.
122. See Brown 1981, 64–65 for the entries of 22 June 1849 and 9 July. See also ibid., 65 for an entry of 7 July.

123. Ibid., 123.
124. See Hueffer 1911, 207 for the opinion R.A.M. Stevenson confided to the author in 1897 that 'all the Fontainbleau [sic] school, all the Impressionists, never did anything but imitate that picture', despite it never being exhibited in France.
125. Brown 1981, 76. See also Brown 1865, 7. Cited in Bennet 2010, 1: 123.
126. See Staley 1973, 27.
127. Minnaert 1954, 135.
128. See Macmillan 1972, 194–195, on how Rossetti believed Tupper held a 'pet theory' about colour contrast. See also Coombs et al. 1986, 33. For William Michael Rossetti's recollection that 'The tendency of his [Tupper's] mind was certainly quite as much scientific as artistic', see Tupper 1897, vii.
129. Tupper 1855, 160.
130. Ibid.
131. Ibid.
132. Coombs et al. 1986, 26.
133. Hunt 1905, 91. Cited in Staley et al. 2004, 25 and in part in Staley 1973, 24.
134. Bennett 1969, 30. This also cites *The Athenaeum* critic on the 'purple shadows' in the painting.
135. See Millais 1899, 119, which states that they worked 'in the morning' from 'eight' until 'seven in the evening'. Cited in Bronkhurst 2006, 1: 147, which identifies the direction of the shadows, and maps out the chronology. On the progress of the painting, see also Millais 1899, 1: 130–132; Hunt 1905, 1: 262; and Bennet 1969, 30. For the suggestion that Hunt's lengthy procedure was necessitated in part by his use of a wet white ground, see also Bennet 1969, 29.
136. See Millais 1899, 1: 130 and Bronkhurst 2006, 1: 147.
137. See also Kemp 1990, 304 on Millais's *Ferdinand Lured by Ariel* of 1849 (later retouched).
138. Twining 1849, 195.
139. See Kemp 1990, 310 on the 'light-toned violet-blues in shadows' used by Boudin, Daubigny, and Corot; and Pomarède 1996, 11 and 30 for the impact of Valenciennes on Corot and his generation.
140. Zola 1868. See Helmholtz 1891, 107 on how surfaces appear uniformly bright under strong sunlight because we cannot detect increases in luminosity above a certain threshold.
141. Thiebaut-Sisson 1900, 20. Cited in Gedo 2010, 51–52. See also Wildenstein 1974–91, 1: 35 for a letter from the painter, A. Dubourg to Eugène Boudin of 2 February 1867, which describes how the work was painted 'in the open air'. Cited in Gedo 2010, 53.
142. See Geffroy 1924, 37 and Mortier 1927, 122.
143. See Wildenstein 1974–91, 1: 35, on how Dubourg witnessed Monet reworking the painting in the studio in February 1867.
144. See Callen 2000, 120, on these 'blue-tinged shadows'.
145. Notes of an interview which took place in Munich in 1910. Cited in Rewald 1973, 210. I am grateful to Anya Hurlbert for drawing this remark to my attention.
146. See Webster and Kay 2007, 34–35.
147. Webster and Leonard 2008, 2818 and Webster 2015, 204.
148. On adaptation in general, see Pearce 2014, 7. On adaptation to blue light, see Winkler et al. 2015, 1.
149. Ibid., 2.
150. Brown 1865, 8. Cited in Staley 1973, 37. See also Bennett 2010, 160–162, on how the work was painted in October 1852 and September and October 1853 and subsequently reworked in 1854 and 1855, after Brown had bought it back from its original owner for this express purpose. This means it is not a 'literal transcript' of the scene as stated in Brown 1865, 7.
151. Brown 1865, 8. On this passage, see Staley 1973, 37. See also Brown 1981, 27, for a diary entry of January 1848 complaining about a lecture in which David Ansted described how 'the colour of the air is blue'.
152. Delacroix 1893, 2: 419. See also Minnaert 1954, 318–319.
153. On this work, see Johnson 2001, 225.

154. See Leclerc 1746, 151–155, for the use of the term, 'accidental colours', to describe the coloured after-images produced by successive contrast. For a summary of Scherffer's ideas about accidental colours, see Boskamp 2011, 49 and 58–62.
155. Ibid., 6.
156. Scherffer 1765, 5–6. See also ibid., 17, for Scherffer's own ideas about the greater refrangibility of blue and their relation to Bouguer's.
157. Ibid., 39.
158. Ibid.
159. Ibid.
160. The present author has performed this experiment with the same unmistakeable result. See also Hardin 1988, 52, for the argument that a colour induced by simultaneous contrast can sometimes generate a negative after-image.
161. Scherffer 1765, 12–14, 16–18, and 24. See De Valois and De Valois 1997, 128–130, for a description of the physiology of successive contrast, and how 'fatigue' is better understood in terms of retinal bleaching and desensitisation supported by neural rebound (whereby a cell tuned to respond to a particular colour surrounded by its opponent complementary will cease to react when that stimulus disappears).
162. Ibid., 33 and 38–39.
163. Ibid., 40, which also argues (mistakenly) that this is proven by the way that objects produce more yellowish after-images under candlelight than normal.
164. Ibid., 41.
165. Ibid., 5.
166. See Mollon 2006, 302, for a similar argument.
167. Scherffer 1765, 41.
168. Ibid.
169. Scherffer 1785. As pointed out in Boskamp 2011, 62.
170. Mazéas 1754, 260. Cited in Beguelin 1769, 38.
171. Ibid.
172. Ibid., 39.
173. Ibid., 38.
174. Ibid., 39.
175. Ibid., 40.
176. Farago et al. 2018, 2: 716. See also da Vinci 1540, f75r and da Vinci 1651, 41. For Leonardo's ideas on how the sea can colours shadows green, see Kemp 1999, 131.
177. Leclerc 1746, 157–158. See also Leclerc 1774, for a letter from Abbé Millot which describes green shadows in an interior space. For a translation, see Baxandall 1995, 29–30.
178. Leclerc 1746, 157.
179. Ibid., 158.
180. Ibid.
181. Ibid., 157–158.
182. Additive mixture is the name given to the process that occurs when beams of light are mixed to combine their power to activate cone cells tuned to the long-, medium-, and short-wavelengths corresponding to red, green, and blue. By this account, 'yellow' light is yellow because it stimulates red- and green-sensitive cones, and 'blue' light is blue because it stimulates blue-sensitive cones. When presented with the yellow and blue light sources together, therefore, all three sets of cones are stimulated, which produces the sensation of white (not green). The laws of subtractive mixture describe the process that operates when pigments are mixed, which combines their individual capacity to absorb light of a particular wavelength. By this account, yellow paint is yellow because it absorbs short-wavelength light and reflects medium- and long-wavelength light, and normal blue pigment is blue because it absorbs long-wavelength light and reflects most short- *and* some medium-wavelength light. It follows that a mixture of yellow and blue pigment will absorb both short- and long-wavelength light and will reflect only (some) medium-wavelength, 'green' light. For a succinct account of the two types of mixture, see Homer 1978, 8–11; and Mollon 2003, 26–29.
183. Beguelin 1769, 29–30.

184. Ibid., 35.
185. Ibid., 30.
186. H.F.T. 1782, 143–146.
187. Valenciennes 1799, 285. Valenciennes also noted here how 'physicists' (by which he probably meant H.F.T.) had tried to explain green shadows in terms of 'weakened light'. See H.F.T. 1782, 206–208, for his claim that a 'green shadow' is cast by a comparatively 'stronger light' or illuminated by a comparatively 'more feeble light', than the light casting or illuminating its relative blue.
188. For the earliest texts to identify the primary colours of light, see Wünsch 1778 and Young 1802. On the history of their discovery, see Crone 1999, 134–140 and Mollon 2003, 13–14 and 26–29.
189. Burtin 1808, 1: 42–43.
190. Beguelin 1769, 30 and 35.
191. H.F.T. 1782, 153–154.
192. Valenciennes 1799 or 1800, 285. Cited in Cavina 2001, xli and Gallo 2017, 183.
193. Valenciennes 1799 or 1800, 285.
194. Goethe 1840, 227.
195. See ibid., xiv.
196. Hunt 1905, 1: 159. See also Glanville 2004, 29, which states that the Pre-Raphaelites admired Goethe's *Theory of Colours*; and Holmes 2015, 701, which points out that Goethe headed a 'series of Romantic heroes' they venerated. It is implicit in the colour circle created by another 'immortal', Isaac Newton, that a mixture of blue and yellow light produces green. See Newton 1704, 114–117.
197. See Bronkhurst 2006, 1: 138 for William Michael Rossetti's record that Hunt made this 'study (in colour) of a cornfield' while at Ewell in August 1849, and for the information (on a label written by Gladys Holman Hunt) that it was painted at 'Ewell, at his uncle's farm' in its initial stage (before the addition of the corn stook in front of the reaper). See also Staley et al. 2004, 66, for the suggestion that its low viewpoint suggests it was painted 'on the spot' because this is consistent with the artist sitting in the cornfield.
198. Beguelin 1769, 35.
199. Ibid.
200. Valenciennes 1799, 287. See Rood 1879, 82, on how green leaves contain a red pigment which can create a yellow colour in sunlight.
201. See Rood 1879, 82–83, for an analysis of the long-wavelength component of the spectrum produced by vegetation.
202. On Walker, see Musson and Robinson 1969, 104–106, 125, 164, and 203 and Ruston 2007, 228–232.
203. Walker 1799, 455.
204. Ibid.
205. See Thompson 1794, 111, for a description of an experiment in which light passed through a filter coloured 'yellow approaching to orange' coloured a shadow illuminated by white light 'blue approaching to green'. See also Thompson 1802, 336, for an experiment in which a 'blue shadow, nearly approaching to green' was produced by illuminating an object with beams of 'deep red light' and 'white or colourless light'.
206. Walker 1799, 456
207. Ibid.

4 Contrast

Eighteenth-Century Theories

The theory that coloured shadows in a natural scene are produced by contrast first appeared in the mid-eighteenth century, probably (as Baxandall noted) in a footnote which Charles-Nicolas Cochin supplied to Jombert's *Méthode pour apprendre le dessein* of 1755.[1] This begins conventionally enough, by attributing shadow colour to a form of scatter enacted by particles of water suspended in the air (described in Chapter 3); but it suddenly changes tack halfway through to refute this argument and identify their cause as contrast. It reads, not a little incoherently:

> When the air is charged with [vapours], shadows are less black, and participate of the colour that the vapours take on from the sunlight illuminating their particles. This colour . . . varies according to the time of day. In the morning, around dawn, it causes shadows to appear blue, and often even violet; towards sunset, shadows are again bluish. It is not that shadows are actually this colour. For all shadows are grey in themselves; that is to say, their colour is extinguished by the privation of light. What causes them to appear bluish is the opposition of golden or red tones, or others which makes the grey look blue.[2]

The first artist to argue that contrast was responsible for the coloured shadows produced by artificial light – using the word 'contrast' in his account – was the illustrator of butterflies and colour theorist, Moses Harris. He argued in *The Natural System of Colours* of *c.* 1770 that when a 'candle' casts a 'shadow' on a 'sheet of white paper' at 'twilight in morning or evening', this will be a 'very fine blue colour', owing to the influence of the 'orange colour' that the 'flame of the candle casts all over the paper, except that part shadowed by the stick'.[3]

There is an element of ambiguity in this account since it does not specify how the simultaneous form of contrast it describes differs from the successive kind described in the previous paragraph. This recounts how, when a 'pair of green spectacles are placed before the eyes, and viewed through for about five minutes, and then taken away', the result is that 'every scene and object will look of a fiery red, opposite to green'.[4] Harris thus effectively collapsed the two kinds of contrast into one, as did several later commentators. Not the least problem this entails is that the colours created by the two processes are different. While, as already mentioned, simultaneous contrast induces the opponent complementary in a field enclosed by one of another colour, the colours of the after-images generated by successive contrast are not easily accommodated by any notion of complementarity.[5]

To make matters worse, Harris specified that the colours generated by contrast (in general) corresponded to those lying opposite one another on the colour circle (Figure 4.1) he based on subtractive mixture.[6] He therefore informed those who wished to know the colour of the shadow cast by warm candlelight to look at the 'prismatic system' (or the colour circle he had loosely derived from the one Newton developed on the basis of his experiments with a prism) which, he claimed, 'will account for this, by informing the reader that blue is the contrast to orange'.[7] Unfortunately, like many of those employed by his successors for the same purpose, a circle of this kind does not indicate quite the same complementary pairings as those generated either by simultaneous or successive contrast, nor indeed those generated by constancy.

Another persistent anomaly in Harris's account – reiterated in many that followed – is that although it characterised the blue colour of the shadow illuminated by neutral daylight as a function of contrast, it characterised the yellow of the shadow lit by the candle in purely physical terms or as wholly immune to any influence from its surround. This is a problem because, although this shadow is not significantly affected by contrast, it is modified by constancy mechanisms, which turn it a slightly dirty colour,

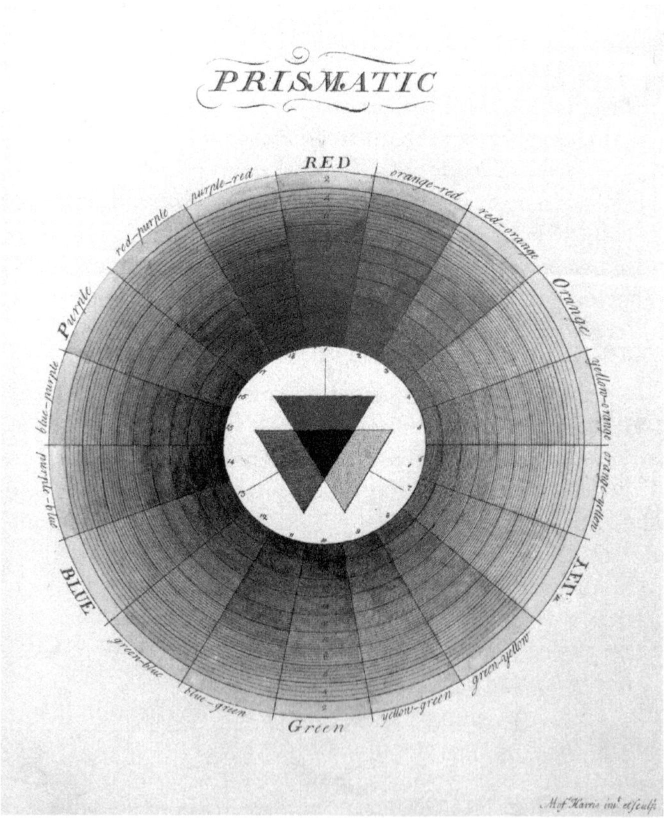

Figure 4.1 Prismatic colour wheel. Plate from Moses Harris, *The Natural System of Colours* (London: Laidler, *c.* 1770). Etching with hand colouring.

Source: Royal Academy of Arts.

as one or two investigators observed later. In effect, then, Harris failed to realise that if he were to explain the colour of both of the shadows he observed, he would need a theory with greater explanatory power.

Cochin had concluded his footnote on contrast and coloured shadows by stating 'these remarks are of a particular concern to the landscape painter'.[8] And Valenciennes, it would seem, concurred with his predecessor, at least to the extent of admitting in the *Elémens* that there was 'another cause' – in addition to reflection – which made 'shadows look bluer than they actually are' at sunrise and sunset, namely 'their opposition to the golden light of the sun at these times of day'. This, he argued, was proven by the fact that when the 'sky is clouded over, the hue of shadows is not as pronounced as it is when it contrasts with the unadulterated light'.[9] Valenciennes also stated that 'Brightly lit objects have a golden tone which contrasts singularly with the bluish shadows visible at sunset.'[10] The process concerned is plainly visible in the foreground of *The Ancient City of Agrigentum* (Figure 3.3), where irregular patches of sunlight bite into the blue shadows cast by the tree and ruin at the left.

Valenciennes was not the first to enunciate the argument that contrast enhanced the blue colour of shadows generated initially by reflection. Rather, Darwin did so in a passage of *The Loves of the Plants* of 1891 concerning shadows coloured by 'blue rays' from the sky, where he stated that:

> [I]f a shadow on a piece of white paper is produced by placing your finger between the paper and a candle in the day light, the shadow will appear very blue; the yellow light of the candle upon the other parts of the paper apparently deepens the blue by its contrast; these colours being opposite to each other.[11]

In the paper on coloured shadows Rumford read in February 1794 and published that year, he described an experiment similar to Harris's in which a 'candle' and a 'beam of daylight' from the North were made to cast two shadows on a 'sheet of very fine white paper'.[12] And like Harris, Rumford found that the shadow 'illuminated by the daylight' was 'yellow', and the other 'illuminated by the light of the heavens' was 'the most beautiful blue it was possible to imagine'.[13] Rather like his predecessor, too, Rumford believed that the 'yellowish hue' of the shadow illuminated by the 'yellow light of the candle' was 'not surprising'; but he was interested in why the shadow lit only by the 'white light of the heavens' appeared blue.[14] Rumford then repeated his experiment using a variety of natural and artificial sources including an Argand lamp, which burned oil and emitted a strong light, and a series of yellow, orange, blue, and amethyst-coloured glass filters to colour one of the light sources. And in each case, he found that the shadow illuminated by the unfiltered light source was opposite in colour to the light casting it.

Rumford then asked whether these induced shadow colours were an 'optical deception, owing to contrast' or 'some effect of the other neighbouring colours upon the eye'.[15] To test this hypothesis, he viewed one such shadow in isolation from its surroundings through a reduction tube or a 'tube about 12 inches long, and near an inch in diameter, lined with black paper', while his assistant repeatedly placed a yellow filter before one of the Argand lamps illuminating the scene and then removed it in quick succession.[16] For Rumford, the shadow remained without 'colour' and underwent no 'change' during this process, while it tuned 'blue' for his assistant whenever the yellow filter was in place.[17] Conversely, when Rumford then took the tube away from his eye and viewed the shadow surrounded by the paper with the 'yellowish hue' this 'acquired' from the filtered light, its 'beautiful blue colour' returned.[18] He therefore concluded that the shadow's blue colour was induced by 'its neighbouring accompaniments', including the 'other shadows rendered really yellow by the effect of the yellow glass', and the 'white paper which had likewise from the same cause acquired a yellowish hue'.[19]

Rumford invoked the notion of contrast in connection with coloured shadows a second time in a paper he composed shortly after the first, 'Conjectures Respecting the Principles of the Harmony of Colours', which was not published until 1802. Here, more particularly, Rumford described how the pairs of coloured shadows produced by two light sources 'afford the most pleasing contrast to the view'.[20] He also stated that the colour of either shadow was the 'complement' to the other.[21] What Rumford meant by this term is that the two colours would produce 'perfect whiteness' (or white light) when mixed, giving as examples the pink and green shadows produced by a neutral light source acting in concert with green and pink light sources, respectively.[22] In effect, then, Rumford assimilated shadow colour here to additive mixture.

To flesh out his contention, Rumford undertook an experiment designed to replicate the perceptual processes at work. This involved placing slips of red and grey paper contiguously on a pink surface, which he believed to be the equivalent of a white surface illuminated by red and white light. Rumford observed that the grey paper appeared bluish green or complementary to the paper in this context and so concluded that the principle responsible for this effect was the same as the one involved in generating coloured shadows. He did not call this principle 'contrast' in this passage, but his earlier use of the term implied as much. He also used the word to describe the related effect, involving the 'harmony of colours', produced when painters 'contrast their colours' to give them a 'great degree of force and brilliancy'.[23] Rumford thus assimilated shadow colour to simultaneous contrast. In this regard, he is not untypical of contemporary investigators, who often concentrated on similarities between phenomena at the expense of the important differences between them.

Rumford only made a passing mention of artists in his earlier paper on coloured shadows, when he mentioned in an aside how he would leave it to 'philosophers, opticians, and painters to determine' the utility of his observations.[24] He did nevertheless mention a phenomenon of particular interest to many later landscape painters by describing how 'two beams of light from different parts of the heavens' could produce coloured shadows.[25] Rumford's ideas did impact on artists. Not least, the landscape painter, Augustus Wall Callcott, made a long abstract of the reprint of Rumford's paper published 'in the Annual Register for 1794' in 1799.[26] And at the head of this, Callcott noted that Rumford had demonstrated how a 'hue or tint is rendered by contrast in appearance very different to what it would be if seen by itself'.[27]

Rumford's ideas also drew the attention of the colourman, George Field, who summarised them in two places in a notebook from 1811, which contained the manuscript for most of his *Chromatics* of 1817.[28] Field first made mention of these ideas in a paragraph citing the summary of Rumford's paper on coloured shadows that the Lunar Man, Samuel Galton, had given in the *Monthly Magazine* for August 1799.[29] This reads:

> Provide two Candles and let the light proceeding from one pass thro' a Coloured glass; Let the other be so far distant that the Shadows [of objects] from the

> coloured and uncoloured lights be equally strong. The Shadow of the latter will be of the colour of the former, and the shadow of the other will be of the colour opposed to it, . . . *with* which it would produce white whichever coloured glass be used.[30]

Field's second reference to Rumford's paper was a summary of the reprint that appeared in *Nicholson's Journal* of June 1797.[31] This is accompanied by a thumbnail sketch of the experimental apparatus Rumford described (Figure 4.2) and is placed beneath the heading, 'Contrast', which clearly indicates how Field understood his predecessor's arguments.[32] It reads:

> Count Rumford by placing a lighted wax candle so that is shone at incidence with the daylight which shone in at a hole in the top of a Window shutter in a darkened room upon a sheet of white paper, found in the shadows of an interposed Ruler towards the daylight Blue and toward the Candle yellow. . . . The Count found similar results from the coloured lights of Candles, & when he rendered the light of the candle, opposed to daylight, Orange by interposing an orange coloured glass, the paper became light crimson & in the shadows blue and orange. . . . He found also that by looking thro' a tube at the centres of the shadows singly that they were apparently colorless [sic], & [an] illusion of vision.[33]

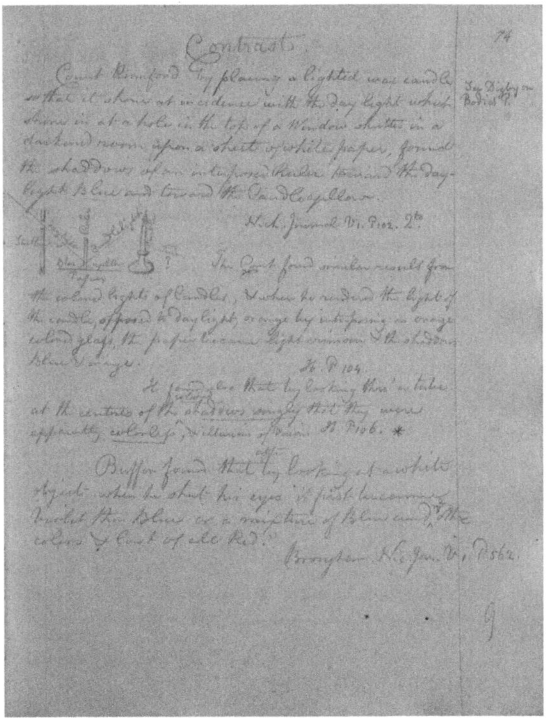

Figure 4.2 Contrast. F74r from George Field, Notebook: *Chromatics or an Essay on the Analogy of Colours* (London: Winsor & Newton, 1811).

Source: Winsor & Newton/author.

Goethe's *Farbenlehre*

Despite the significance of Rumford's paper, Goethe only accorded it a passing mention in the second historical volume of the *Farbenlehre*. This may be because he considered its claim that shadow colour was an optical deception to be 'blasphemy'.[34] Or it may be because Lichtenberg alerted him to the possibility that contrast was responsible for producing coloured shadows in the letter he sent him in October 1793. Here, at all events, Lichtenberg seemed to suggest that a process akin to (simultaneous) contrast could do so, when he added to Goethe's observation that shadows illuminated by a neutral or 'overcast sky' can appear 'blue' by suggesting that this happened when they had a 'yellow context'.[35] Lichtenberg also invoked (successive) contrast when he suggested that 'coloured shadows' belonged to the class of 'accidental colours' Buffon had discovered, which normally appear 'when one looks for a long time through a red glass and removes it suddenly from before the eyes, so that objects appear for a moment green'.[36]

Goethe nonetheless derived his arguments from an experiment – like Rumford's – which investigated the 'beautiful blue' shadow produced on a sheet of white paper by a 'lighted candle' and 'declining daylight'.[37] And just as his predecessor had done, Goethe argued that the colour of the shadow cast by a light source passed through a 'coloured glass' is 'complemental' to this, if the other light source is neutral.[38] He also followed Rumford when he suggested that the 'shadow' in such a situation is always 'opposite' in 'hue' to that of the 'surface' on which it is 'thrown'.[39]

Goethe made a more original claim, however, when he suggested that 'we can only persuade ourselves by some attention that the white paper acts as a reddish yellow' in inducing blue shadow.[40] What Goethe meant is that it is not immediately apparent that the blue of the shadow in such a situation is a contrast effect, until we attend closely to the white paper around it, when the warmth this acquires from the candlelight becomes noticeable. What Goethe hit on is that although we do not see the paper as yellow under normal circumstances, or as we scan over the scene, we can do so by fixing the eyes on it. What actually happens here is that doing so inhibits the processes of comparison on which constancy depends and thereby impairs its ability to make the paper look white. However, to support his claim that it was the colour of the surroundings that produced the shadow's colour by dint of contrast, Goethe greatly exaggerated its saturation in the diagram illustrating his experiment in the *Farbenlehre* (Figure 4.3). And perhaps tellingly, its colour is significantly paler in at least one of the preliminary studies he made for the plate.[41]

Goethe also persuaded himself that the theory of contrast could make sense of observations he had made earlier, including some he had noted in the diary he kept of his trip to the Harz mountains in 1777. It seemed, for example, to explain the fact 'shadows tending to violet' could be seen on the snow on the Brocken 'during the day', when this acquired a 'yellowish hue' from the sun.[42] Contrast seemed to explain as well how these shadows turned 'blue' as the 'sun was sinking', and the surrounding snow became 'yellow deepening to orange'.[43] And ostensibly, it explained why the fact that the 'shadow colour changed to a green' when the 'sun was about to set' and 'began to diffuse a most beautiful red colour' over the snow.[44] By the same token, Goethe believed that contrast could elucidate the unusual observation, originally made by William Halley and reported by Newton in the *Opticks*, that when 'sun light' shines into a diving bell 'under water' and 'everything is seen in a red light', the 'shadows appear green'.[45]

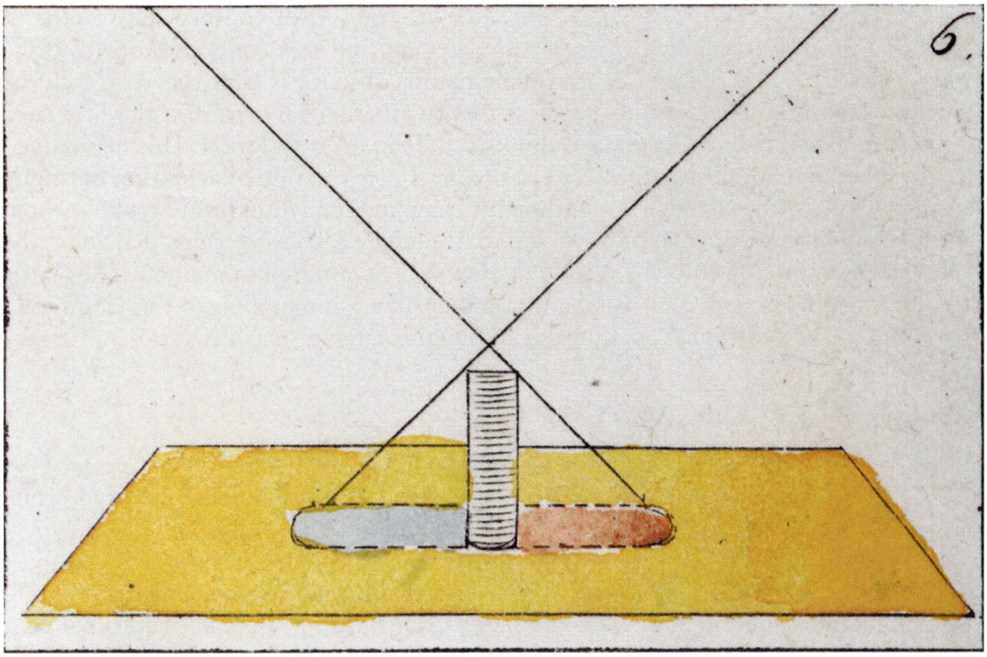

Figure 4.3 Coloured shadows. Fig. 6, Plate 1 from Johann Wolfgang von Goethe, *Zur Farbenlehre* (Tübingen: Cotta, 1810).

Source: Author.

What Goethe meant by 'contrast' remains unclear, however.[46] For one thing, he did not distinguish between successive and simultaneous contrast in the section of the *Farbenlehre* on coloured shadows, nor in the immediately preceding section, where he treated a range of contrast phenomena at length.[47]

More importantly, some aspects of Goethe's theory of the role played by contrast in generating coloured shadows are quite esoteric. One of these is the idea that an orange shadow will turn red as result of a 'deepening' process called 'augmentation', which Goethe regarded as a function of how colour is produced by the mixture of light and dark.[48] This notion emerges most explicitly in a description of an experiment in which a 'white surface' lit by the 'full moon' and a 'candle' produces a 'double shadow'.[49] At first, Goethe writes, the shadow 'cast by the candle' was 'the most beautiful blue', while the other, 'cast by the moon', was a 'powerful red-yellow'.[50] However, he continued, as the 'eye' dwelled 'on these colours', the 'increasing red in the one' changed the originally blue 'contrast' colour of its sister to a 'kind of sea-green'.[51] While this account is undoubtedly poetic, the process of 'augmentation' it describes is nonetheless fanciful.

A yet more abstruse dimension of Goethe's theory is implicit in his claim that the colours 'evoked' by 'contrast', which lie 'diametrically' opposite one another on the 'chromatic circle', will 'reciprocally evoke each other in the eye.'[52] 'Thus', he explained, 'yellow demands purple; orange, blue; red, green; and vice versa', and 'all intermediate gradations reciprocally evoke each other; the simpler colour demanding the compound, and vice versa.'[53] Elsewhere, the same conception of contrast led Goethe to assert that the 'eye especially demands completeness, and seeks to eke out the colorific circle in itself.'[54] The reason these arguments are obscure is that they can only be understood in terms of the metaphysical theory Goethe elaborated in the foreword to the *Farbenlehre*, which Eastlake did not include in his translation. This maintained that the eye spontaneously generates chromatic effects capable of achieving harmony because it is a component of a continually changing, active natural system, geared towards organising its parts into an 'animate unity'.[55] The eye thus 'demands' the complementary of any colour it is presented with in an attempt to maintain chromatic equilibrium. Phenomena such as adaptation may give some support to this claim indirectly. In Goethe's formulation, however, it remains largely specious.

Colour Theory in Runge and Others

Runge mentioned the red shadows visible though a diving bell in his essay of *c.* 1806–10, 'Gespräche über Analogie der Farben und Töne', where he declared, using musical analogies:

> When the sun sets and it is red. . . . The sky on either side is green, just as it would be blue if the setting sun were orange. So the sunbeam puts itself into contrast with the pressure in the air. The same thing happens when one descends into the deep sea in a diving bell. At a great depth of 100 fathoms . . . the light that falls on one's hand is a deep, splendid red, while the shadow is an equally beautiful green.[56]

Runge referred to Newton many times in his writings.[57] He did not speak English, however, so it is unclear whether he learned of Halley's observations on underwater green shadows himself, or from Goethe, with whom he corresponded about colour for several years, beginning in July 1806, when the poet then sent him a letter inviting him to discuss colour.[58] Later, they also exchanged publications. Runge sent Goethe the manuscript of the *Farben-Kugel* in September 1809 and a copy of the book itself in February 1810.[59] And a few months later, he received a copy of the *Farbenlehre*.[60]

It is also unclear whether Goethe stimulated Runge's interest in contrast or vice versa.[61] Most likely, they encouraged each other. The painter sent a letter to Goethe in early December 1806 which certainly suggests he was incorporating ideas about contrast they had discussed into his recent works. Referring to the paintings (including *Morning*) he intended to make from his four engravings of the *Times of Day*, Runge told Goethe:

> Maybe these four paintings would appeal more to you by the greater unity of the whole as much as through the contrast of tones. It is this contrast of the tones that I would like to have told you more about, and also some sketches and general ideas about *Ossian* . . . which perhaps mostly clear the way towards the real appearance of the colour theory already touched on. . . . I should be really pleased if I should find this phenomenon touched on in your Colour Theory. I have in Hamburg a painting which I have started on, which is the *Flight to Egypt*. I should like to send you the sketches of it so as to acquaint you with it at least.[62]

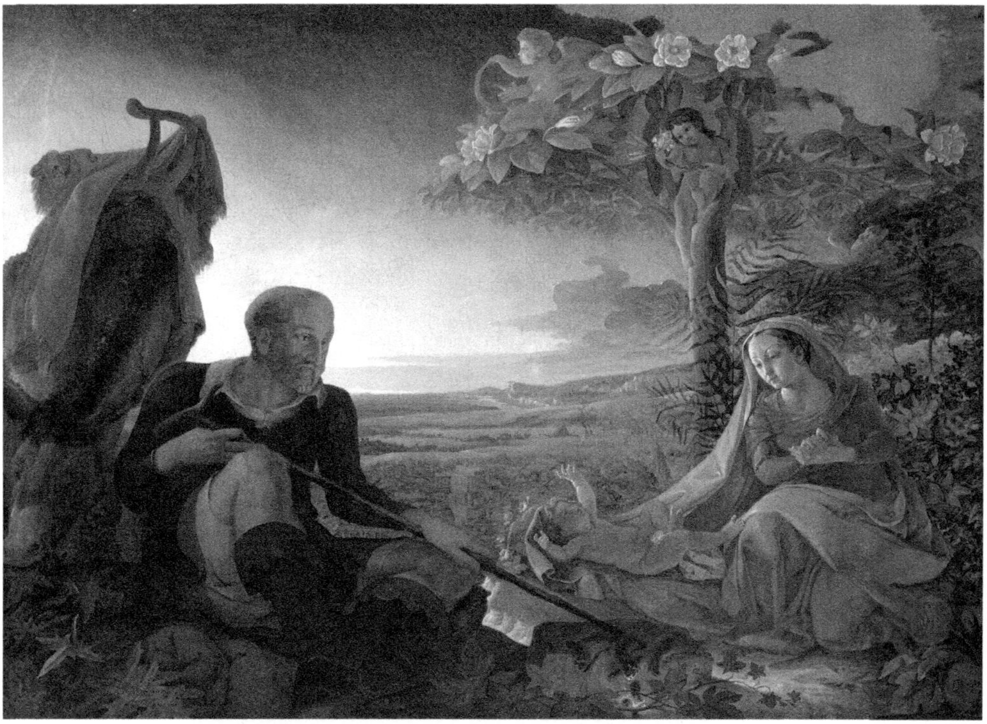

Figure 4.4 Philipp Otto Runge, *Rest on the Flight to Egypt*, 1805–06. Oil on canvas, 96.5 × 129.5 cm. Hamburg, Hamburger Kunsthalle, inv. no. HK-1004.

Source: Hamburger Kunsthalle/Bridgeman Images.

The more particular interest in coloured shadows Runge shared with Goethe finds expression in the lines of 'Gespräche über Analogie der Farben und Töne' that follows immediately from those cited earlier. These describe how 'when one lies on the grass and holds a hand up into the pure air to stop the sun shining into one's eyes, the illuminated side becomes beautifully yellowish orange, but the shadow an equally beautiful bluish violet.'[63] Runge incorporated this gesture into his paintings, *Rest on the Flight to Egypt* of 1805–06 (Figure 4.4) and *Morning (second version)* of 1808–09 (Figure 4.5). But he did not depict the colours in the hands in either painting quite as he described them in his essay. Instead, close examination of the first painting reveals yellow highlights in the Christ child's upraised left hand and red mid-tones which cause the physically grey shadows next to them to turn slightly green. In the later work, the shadows in the hands assume a bluish cast, but only because the flesh colour is applied so thinly over the dark layer below as to be semi-transparent; while at the same time they are tinged with green reflections from the surrounding verdure.

Being as they are, the colours of light and shadow in these works differ significantly from the 'contrasts' lying 'opposite one another' on the colour 'circle', which lay out the 'tone of the light' and the 'colour of the shadow' in a single view, as Runge stated in his essay 'Weitere Ausführung' of 1809.[64] Instead, these relationships are subordinated

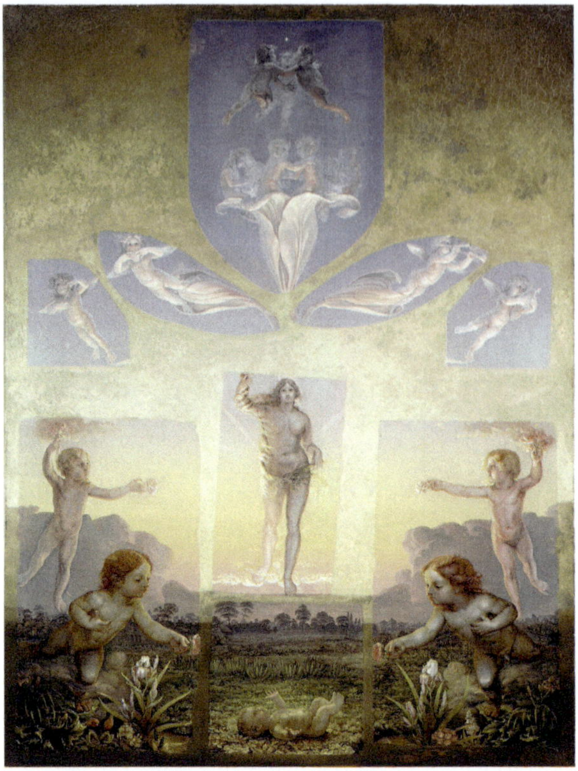

Figure 4.5 Philipp Otto Runge, *Morning (second version)*, 1808–09. Oil on canvas, 152 × 113 cm. Hamburg, Hamburger Kunsthalle, gift of Conrad Meißner, inv. no. HK-1022.

Source: Hamburger Kunsthalle/Bridgeman Images.

in *Rest on the Flight to Egypt* to what the artist described in the *Farben-Kugel* as the 'harmonic contrast' of orange and green.[65] While in *Morning*, they are subservient to the 'harmony' the same text discerns in the combination of red and green, and blue and orange – at least if pink can stand in for red and gold for orange.[66] The fact Runge equated harmony and contrast could also indicate his agreement with Goethe, who mentioned in several places in the *Farbenlehre* how 'contrasts' and 'complemental' colours, including those manifested in after-images and coloured shadows, created 'harmony' by virtue of producing 'completeness'.[67]

Ideas about contrast may have played a role in Friedrich's painting, since (as already mentioned) the artist knew Runge. He also knew Goethe, who visited him in Dresden in September 1810, a courtesy the painter reciprocated when he visited the poet in Jena the following year.[68] The warm–cool contrasts of *Hill and Ploughed Field Near Dresden* are at all events similar to the orange and blue contrasts Runge mentioned in 'Gespräche über Analogie der Farben und Töne', as well as being explicable by physical theory.

Carus met Goethe once in 1821.[69] Before this, however, and afterwards, they regularly corresponded about scientific matters; and Carus sent Goethe four of his letters on landscape painting.[70] Such was Carus's esteem for Goethe that he used a letter the poet sent him in April 1822 as an introduction to the published edition of these letters, *Neun Briefe über Landschaftsmalerei* of 1831.[71] This text also manifests a debt to the *Farbenlehre* in its use of the term 'physiological colours', which Goethe had employed for describing the colours produced by contrast.[72] And like Goethe, Carus believed that contrast coloured shadows. In a letter of December 1822, for example, he described how, one evening, 'Fresh snow decks the spruces and pines, but in shadow looks violet, contrasting with the flush of sunset in the air'.[73] The distribution of blue shadows in *A River Cruise on the Elbe River Near Dresden* of 1827 (Figure 4.6) indicates that Carus envisaged them as physical in origin or lit by skylight. This painting nevertheless captures beautifully how the yellow colour of the oarsman's waistcoat under warm sunlight enhances the colour of the shadows next to it, particularly of the shadow immediately to its right.

Another painter whose depiction of coloured shadows was indebted to Goethe was Kopisch. While passing through Weimar in December 1929, he wrote to his hero to request a rendezvous.[74] And although the painter described himself in his note as

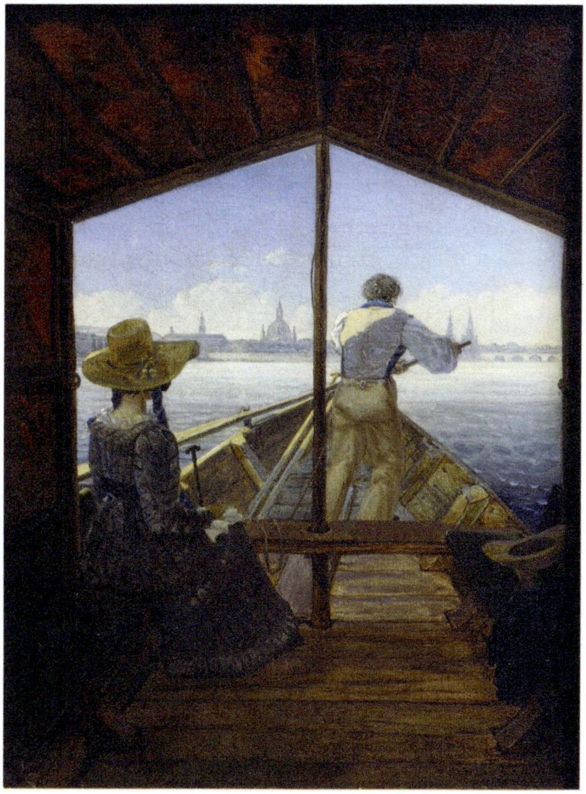

Figure 4.6 Carl Gustav Carus, *A River Cruise on the Elbe River Near Dresden (A Morning on the Elbe River)*, 1827. Oil on canvas, 29 × 22 cm. Düsseldorf, Kunstpalast, inv. no. M 130.

Source: Kunstpalast/Paul Smith.

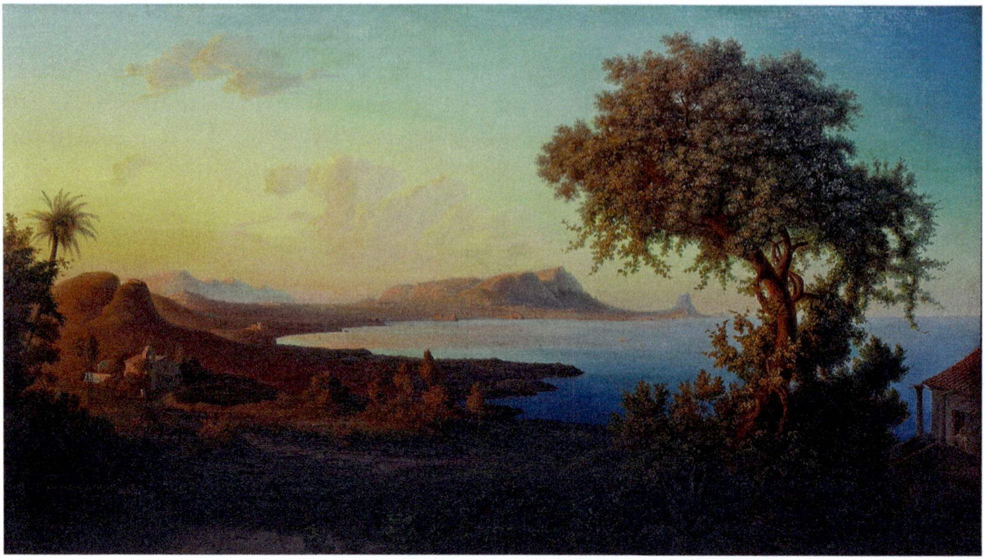

Figure 4.7 August Kopisch, *Cape Zaffarana*, 1840. Oil on canvas, 62 × 110 cm. Bad Homburg, Verwaltung der Staatlichen Schlösser und Gärten Hessen, Schloss Bad Homburg v. d. Höhe.

Source: Renate J. Deckers-Matzko.

a 'newcomer to poetry', it is hard to believe that he did not know the *Farbenlehre*. It is likely, then, that this text informed the spectacular green shadows Kopisch depicted in *Cape Zaffarana* of 1840 (Figure 4.7).[75] These shadows are most clearly visible in the walls of the small church at the left, where they are clearly generated by the sunlit red areas alongside them. This work was evidently painted from a high vantage point, so it could have recorded shadows Kopisch actually saw on the spot. The fact, however, that the reds and greens in the shadows are repeated in tiny spots on the trees surrounding the church indicates that the artist was interested in exploiting contrast effects synthetically as well. It does so, more specifically, in a manner consistent with Goethe's idea that contrasting colours create harmony because they achieve completeness. It chimes in especially with remarks Goethe made after mentioning the green shadows he had seen on the Broken and those that appear underwater in which he claimed that the 'very same phenomenon which I observed on a high mountain is presented to others in the depths of the sea, and thus Nature throughout is in harmony with herself.'[76]

A lesser well-known painter with an interest in coloured shadows who corresponded with Goethe was Jacob Roux. Goethe initiated their correspondence in January 1815, although the artist had made an etching of the poet's garden at Weimar around 1810.[77] Roux had studied mathematics at University in Jena and subsequently anatomy, before he took up a position as Professor of Anatomical Drawing at Heidelberg University in 1818, an act which thwarted Goethe's to involve him in setting up a new art school in Jena.[78] Given this close association, it is no surprise that the discussion of coloured shadows in the third and last volume of Roux's treatise, *Die Farbe*, published under the title *Entdeckungen aus dem Gebiete physikalischer Farbenlehre* in 1829, was much indebted to Goethe's ideas about contrast and particularly to his notion that the eye 'demanded' the complementary of the colour presented to it.

This debt is evident in Roux's account of experiments demonstrating the 'orange and blue shadows evoked at dawn by candlelight and daylight', which contests that 'one of the shadows [is] coloured by the demanding colour orange, the other appears as the blue demanded colour'.[79] Unlike his predecessor, however, Roux implied that it was the light sources themselves, and not their surrounds, that coloured shadows. He also suggested that their capacity to do so was enhanced by their power to intensify each other's colour by contrast. Thus, he claimed that 'Candlelight that shines a pale yellow at night appears a more colourful yellow when it is surrounded by daylight' and that 'When there is a candle burning in the room, the eye perceives the grey daylight reflected from the sky as bluish-grey'.[80]

Roux's thinking became more idiosyncratic when he discussed the effect of the 'reddish colour' that supposedly emerged when two 'coloured beams of light, one generated by a candle and the other by daylight' are made to 'cross over one another on a white surface' at the 'point where they are superimposed'.[81] This colour featured particularly in the descriptions Roux gave of a series of experiments involving the box and set of coloured filters he made for creating 'coloured shadows' (Figure 4.8).[82] So, for example, when Roux admitted 'candlelight' and 'daylight' through the box's apertures

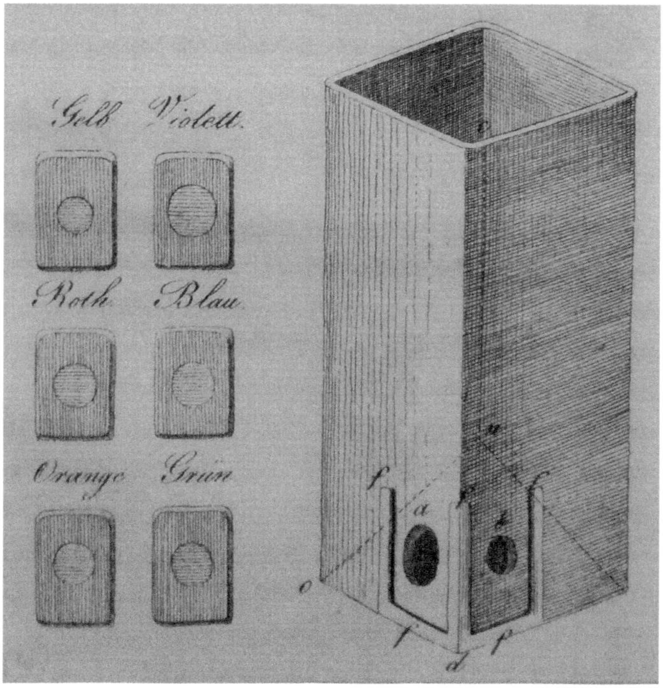

Figure 4.8 Box and coloured filters for creating coloured shadows. Figs III and IV from Jacob Roux, *Entdeckungen aus dem Gebiete physikalischer Farbenlehre* (Heidelberg: C.F. Winter, 1829). Worcester College, Oxford.

at 'dawn and dusk' and positioned the candle so that its 'fiery orange' lay over the 'greyish blue' of the daylight, a 'reddish colour' appeared where the two sources met.[83]

This did not 'exhibit such a beautiful redness', however, as the colour evoked in another experiment, which demonstrated Roux's predilection for tracing the sometimes bewildering sequences of knock-on effects ostensibly produced by the interaction of light sources.[84] The experiment in question involved placing a 'thin stick' in the 'centre of the reddish glow' created where light sources of the kind described in the previous experiment overlapped to create two new 'shadows', one a 'fiery orange' coloured by the candlelight and enhanced by its 'pure, pure blue' neighbour, and one of 'beautiful blue demanded colour' which the orange reciprocally induced – but only after the candlelight had first 'raised' its objectively greyish-blue colour (as 'daylight') 'to blue'.[85] The particular red visible under these circumstances, Roux stated, was nothing other than a 'primary colour' by virtue of standing out against a 'warmer orange' and a 'colder blue'.[86] Roux also described a second series of experiments using filtered 'sunlight' in place of candlelight and 'greyish blue' daylight, which confirmed the results of the first.[87]

For all Roux's certainty about his observations, it is difficult to credit them. This is particularly true of the complex effects produced by sequences of events. But even his relatively straightforward claim that simply superimposing blue and orange light sources creates a red colour is suspect, since the additive mixture of orange and blue light produces an almost neutral, only slightly magenta, resultant.

Roux did not produce any paintings which match his ideas perfectly. However, his *View Heidelberg Castle and the City From the East* of 1816 (Figure 4.9) does depict a

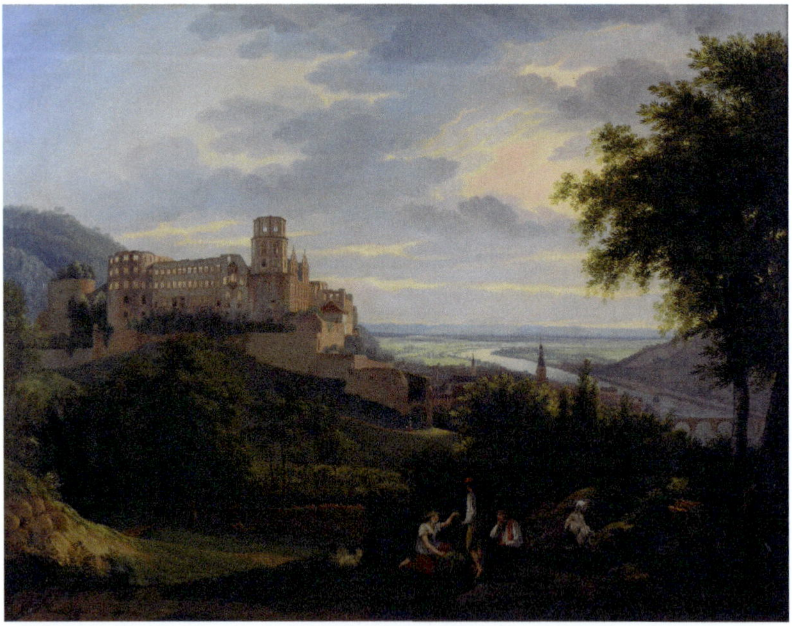

Figure 4.9 Jacob Wilhelm Roux, *View of Heidelberg Castle and the City From the East*, 1816. Oil on canvas, 65.5 × 83.5 cm. Schweinfurt, Museum Georg Schäfer, inv. no. MGS 2340.

Source: Museum Georg Schäfer, Schweinfurt.

landscape illuminated by the reddish setting sun as it emerges from behind the clouds. And perhaps the walls of the castle facing away from the sun look just a little cool by comparison with those that face it.

Goethe's ideas about the 'demanded' nature of shadow colours re-emerged in many places in later years. Their manifestation in Guido Schreiber's *Die Farbenlehre* of 1868 is noteworthy, however, since this contained colour illustrations (Figure 4.10) of experiments concerning these. The first shows the familiar experiment demonstrating the shadows produced by a combination of 'lamplight or gas flame' and the natural light of 'dusk'.[88] And like Goethe, Schreiber maintained that while the 'yellow-orange' or 'roughly dark chrome yellow' shadow took its colour directly from the 'lamplight' illuminating it, the 'pure blue' shadow lit by daylight assumed the 'complementary' colour 'demanded' by the yellow light illuminating the 'white surface' surrounding it.[89] So, too, Schreiber claimed, the 'sea-green' shadows shown in the second illustration he generated by passing lamplight through a 'red' filter were 'complementary colours' of the same kind.[90]

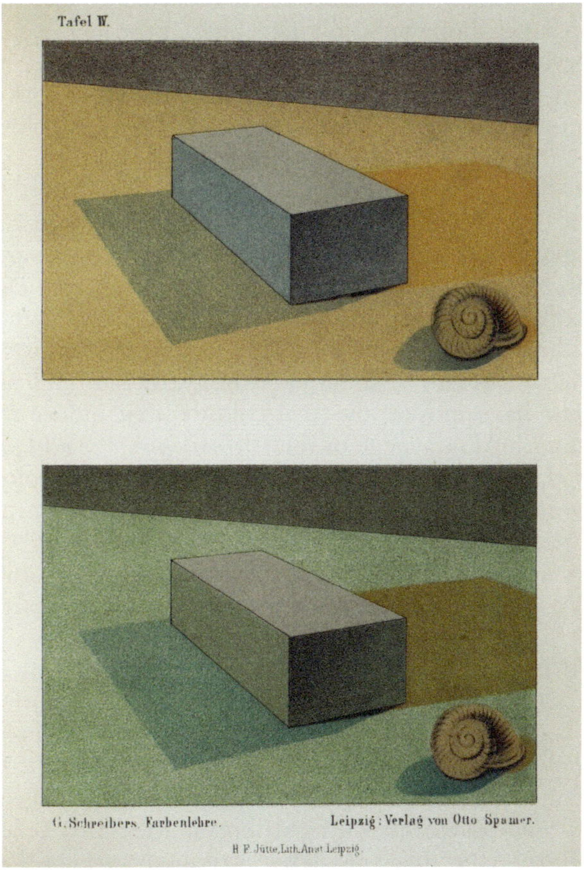

Figure 4.10 Coloured shadows. Plate IV from Guido Schreiber, *Die Farbenlehre* (Leipzig: Otto Spamer, 1868).

Source: Author.

The Italian Perspective

The Italian painter and art historian, Giuseppe Bossi, discussed the role played by contrast in generating coloured shadows in a paper he delivered in 1814 and published in 1821, 'Saggio di Richerche intorno all'armonia cromatica'.[91] Here, more particularly, Bossi described two experiments demonstrating how contrast coloured shadows with the hue 'consequent' or complementary to the 'real colour' the surface on which they fell.[92] The first involved viewing objects with the aid of a 'coloured glass inclined at about forty five degrees on the optical axis'.[93] Bossi observed that when he looked through the glass, surfaces that were 'neutral' in colour and illuminated 'took on the colour of the glass'.[94] But, when he observed the 'reflections' on the surface of the glass emanating from neutral areas in shadow, he found that these 'participated in the colour consequent to that of the glass'.[95] This experiment is perhaps a little suspect, however, since it runs counter to the fact Monge discovered that some coloured objects (particularly red and yellow) look white through coloured glass.[96]

No such misgivings attach to Bossi's second experiment, whose results are confirmed by numerous other investigators. This involved casting a shadow into an area on a sheet of paper illuminated by light filtered through a small pane of green glass.[97] Bossi observed that this 'shaded' portion of the paper 'did not look green' but instead 'exhibited a red so pronounced as to equal or win out over the strength of real colour'.[98] Shadow colour was thus demonstrably contrast colour as far as he was concerned.[99]

Although Bossi did not declare a debt to earlier investigators for this theory of coloured shadows, he did acknowledge that he drew his more general theory that 'each real colour places the sensation of its opposite colour in our eye' from Buffon and Scherffer (who had mentioned coloured shadows) and from Darwin and Goethe (who had both explained coloured shadows as contrast phenomena).[100] Bossi was also indebted to Goethe, having read an Italian translation of his writings on colour made by Count Pietro Moscati.[101] In all probability, therefore, Bossi's statement that 'consequent colours' lie 'opposite' the 'real colours' that induce them on the 'chromatic circle' indicates his knowledge of the *Farbenlehre*.[102] And Goethe undoubtedly lay behind Bossi's claim that the phenomenon of 'consequent' colours 'demonstrates the need of nature to bring the organ back into balance with a motion contrary to the real colour impression'.[103] Bossi also cited his predecessor in a passage where he suggested that 'sensations' of 'consequent' colours could 'occur during the sensation of the real colour', as opposed to afterwards, which effectively describes simultaneous contrast.[104]

Bossi mentioned Goethe once more again in a passage which set out the theory that the 'sensation of the colours is obtained by the impression of a given motion in the organ' which is 'different in the same fibre or papilla'.[105] The origin of this was not Goethe's *Farbenlehre*, however, but the *Indagine fisica sui colori* published by his compatriot Giambatista Venturi in 1800. In the section of this entitled 'Dei colori immaginari', Venturi had sketched out how fibres in the retina generate sensations of colour, including 'imaginary and consecutive' colours, through the movements they make in different directions in response to light.[106] Bossi thus synthesised Venturi's ideas with Goethe's when he argued that these papillae generate 'contrasting colours' and colours in 'shadow', when they move to remain 'in balance' with one another.[107]

Perhaps surprisingly, Bossi hardly mentioned painting in his essay, save for noting in his conclusion how the Renaissance masters had employed *cangiante* colour. Here, he argued with respect to this technique that 'practice had taught our elders what we have learned through the support of theory', namely to place 'violet shadows in yellow' draperies, 'orange shadows in pale blue' draperies, and 'red shadows in green' draperies.[108] But despite making this perspicacious observation, there is no evidence to suggest that Bossi applied it in his own work.

Bossi's remark was nevertheless reiterated by another painter and art historian, Girolamo Calvi, in *Della norma che per dipingere le ombre* of 1842. This mentioned how, in representing 'shot silk', Renaissance and Baroque painters effectively placed 'consequent' rather than merely 'arbitrary' colours in the shadows.[109] Indeed, Calvi cited Bossi no fewer than 11 times. So it is fair to conclude that his predecessor's ideas played a major part in Calvi's refutation of the 'absurd' argument advanced by Leonardo, Buffon, Priestley, among others, that shadows were coloured by 'reflections' from the sky.[110] Bossi's ideas, alongside those of Rumford, Chevreul, and Brewster (and supposedly Scherffer and Meusnier), also informed Calvi's rival claim that contrast was responsible for 'complementary' shadow colour, and his discussion of the different kinds of 'complementary' colours produced by 'successive' and 'simultaneous' contrast.[111] Calvi also discussed at some length the argument made by Bossi, Scherffer, and Chevreul (among others) that contrast colours could be predicted by the colour circle.[112]

Although erudite, Calvi's book was largely unremarkable, save for the ingenious experiment he devised for identifying the complementary colours of shadows. As the illustration shows (Figure 4.11), this involved viewing a 'white card' placed in shadow (at 2) through a small hole in an illuminated 'coloured' card (at 1).[113] The result, Calvi asserted, is that the white card will exhibit the colour 'complementary' to that of the

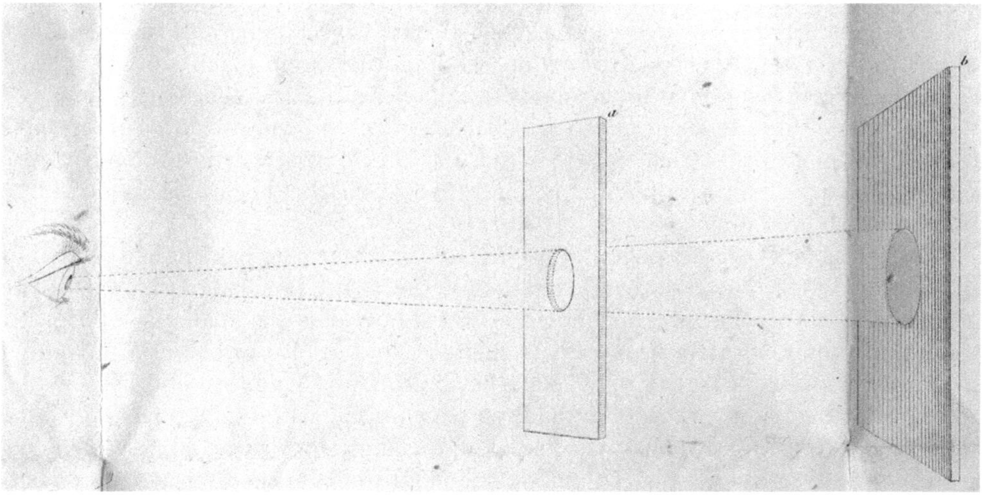

Figure 4.11 Experiment with coloured shadows. Fig. 6 from Girolamo Luigi Calvi, *Della norma che per dipingere le ombre* (Milan: Luigi di Giacomo Pirola, 1842). ND1489 C35, Special Collections, Robert B. Haas Family Arts Library, Yale University.

coloured card or its notional shadow colour.[114] Calvi then repeated the experiment with a series of differently coloured cards to identify their complementary shadow colours. And afterwards he performed another series using coloured cards in place of the white card to find the 'shadow' colours induced in different kinds of 'coloured body' by others nearby.[115] Whether reliable or not, these experiments are certainly testimony to a deep interest in the theory of coloured shadows in nineteenth-century Italy.

George Field and the Pre-Raphaelites

Although Field was indebted to Rumford for his understanding of coloured shadows, he also devised and performed his own experiments and drew his own eccentric conclusions from them. He mentioned one such experiment on the 'Colour of Shadows' on a page of his 1811 notebook headed 'Contrast of Colours', which states cryptically: 'If light be of any colour the shadow cast by such light will always exhibit the contrary colour. Thus, if a ruler be held across the prismatic spectrum, will be demonstrated universally'.[116] Fortunately, Field rehearsed the same experiment in greater detail in the section of his *Chromatography* of 1835 headed 'Colours of Shadows &c.'.[117] This described how holding a ruler in the path of a 'solar bow' (or spectrum) projected on a white screen 'in a room nearly darkened' produced a 'inverse spectrum' in its shadow, whose 'perfect contrasts' corresponded to the original 'colour for colour'.[118] Field did not make it explicit here, however, that a little daylight must be admitted into the room for this experiment to work (as Charles Bourgeois did in a later account of a similar experiment).[119]

On the page of his notebook following the account of the inverted prism, Field made an extract of an experiment with an inverted spectrum which Thomas Young had described in *A Course of Lectures on Natural Philosophy and the Mechanical Arts* of 1807. Under the heading, 'Contrast' and accompanied by a copy of Young's illustration, this explained how holding a slightly convex lens placed together with a sheet of glass between the eye and the 'edge of dark object' generates two sequences of colours, one which 'begins from white' and another 'from black', in which 'each colour is the contrast of that of the opposite half of the ring'.[120] What Young described was in fact an 'interference' effect, which he took as a demonstration of its wave form.[121] In spite of this, Field recorded that the 'phenomenon' was 'probably a true illustration of the contrast of light and Shade in colours' that he had 'spoken of at the bottom of the preceding page'.[122]

Field included a revised version of his inverted spectrum experiment in a section of the second (1845) edition of the *Chromatics* on the 'colours of shadows'.[123] In this, he also concluded that the experiment demonstrated how 'shade is in all cases a contrast to light, not only in effect or power, as chiaroscuro, but also in colour'.[124] Working from this premise, he argued that Leonardo, Buffon, and Rumford had been wrong to attribute the colour of the 'blue shadows which occur with the orange light of the rising and setting sun' to 'blue reflected from the sky', since this colour was 'merely complementary to the orange, or golden colour of the light'.[125] By way of proof, he offered the evidence that if 'any colour be given to the sun's light, by passing it through coloured glass into a darkened chamber, the shadows of such light will always be of the colour which contrasts it, notwithstanding the reflex blue of the sky'.[126] And in both the 1835 and 1841 editions of the *Chromatography*, Field described how his close acquaintance, Thomas Girtin, had 'delighted in' a similar 'effect of sunlight and shadow', wherein a 'deep orange' or 'red' sun generated 'purple' in the shadows.[127]

Since Field makes no mention of any part played by a second light source in these accounts either, it must be assumed he did not grasp its role in generating coloured shadows. Nor does he seem to have believed that a coloured light source colours a shadow indirectly, by first colouring its surround. But these lacunae did not prevent many artists, including the Norwich painter, and subscriber to the *Chromatography*, Joseph Clover, from embracing Field's ideas.[128] He wrote on the page of a sketch-book headed 'Mr Field' and dated 'June 1835': 'The colour of the shadows is always complementary to that of light'.[129]

Hunt told F.G. Stephens in an unpublished letter of October 1879 that he had read the 1841 edition of the *Chromatography*.[130] It is possible, therefore, that he was interested in Field's ideas about coloured shadows almost from the outset of his career as a Pre-Raphaelite. Field's influence is arguably manifested in *Our English Coasts* of 1852 (Figure 4.12), a 'sunlight picture' which Hunt executed according to 'Pre-Raphaelite principles', by synthesising several views of a beauty spot between Hastings and Fairlight, which he made out of doors between early August and late November.[131] At all events, the fleeces of the sheep in this work are blue–violet in the shadows opposite the warm setting sun, exactly as Field's theory of contrast predicted.[132] The colour

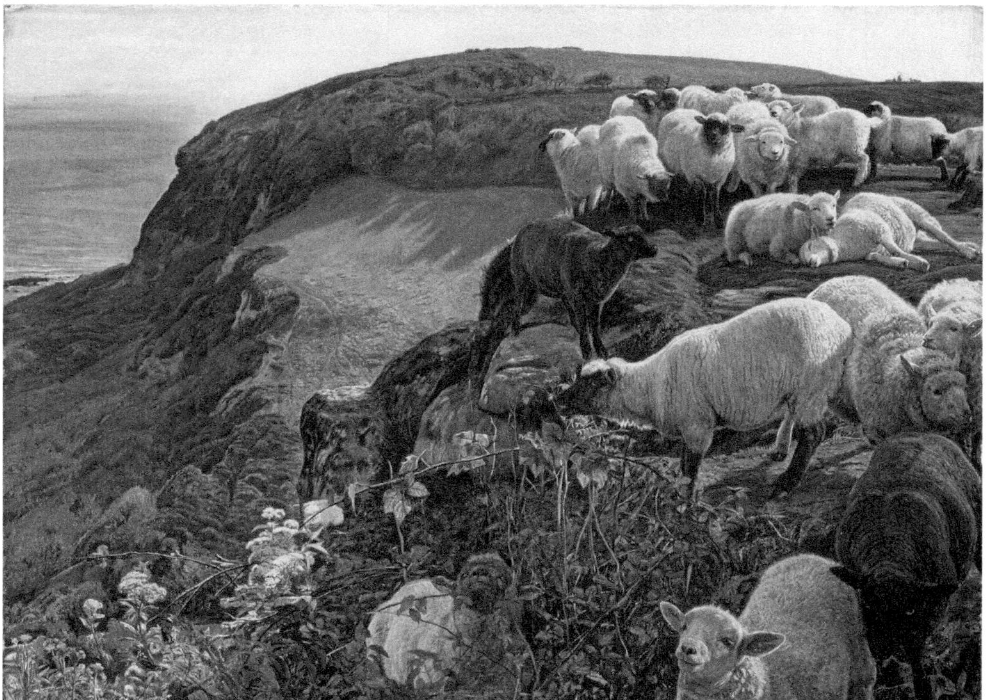

Figure 4.12 William Holman Hunt, *Our English Coasts ('Strayed Sheep')*, 1852. Oil on canvas, 43.2 × 58.4 cm. London, Tate Britain, ref. no. N05665.

Source: © Tate.

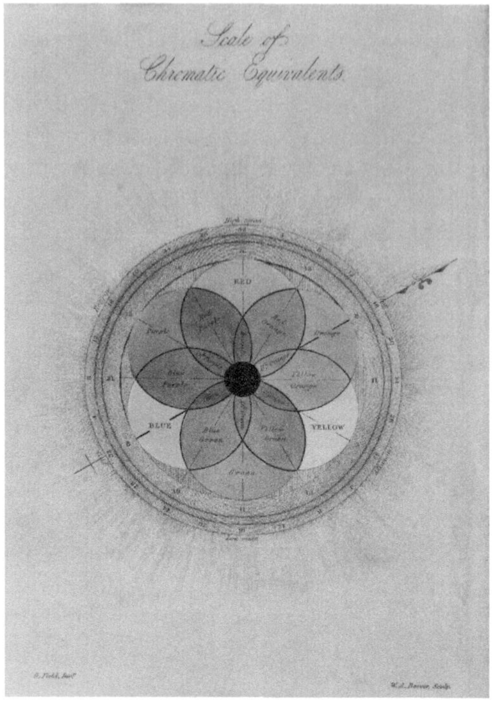

Figure 4.13 Scale of Chromatic Equivalents. Frontispiece to George Field, *Chromatography*. 2nd edn (London: Tilt and Bogue, 1841).

Source: Author.

relationships in this work correspond as well to those set out in many of the colour diagrams Field made and notably the *Scale of Chromatic Equivalents* (Figure 4.13) that served as the frontispiece to the 1841 edition of the *Chromatography*, which characterised 'orange and blue' as 'antagonist or opposite colours'.[133] It must be admitted, however, that Hunt's combination of warm and cool colours is closely similar to the pairing produced by the physical effects of sunlight and skylight at dusk according to Leonardo.

Simultaneous Contrast in France

Claude Antoine Prieur was the first French scientist to describe the 'effect of viewing two differently coloured substances simultaneously, when they are close together' in his important essay, 'Considérations sur les couleurs et sur plusieurs de leurs apparences' of 1805.[134] Here, he also noted that 'painters know' this effect 'very well'.[135] Chevreul nevertheless helped to dispel confusion over the difference between 'simultaneous' and 'successive' contrast by clearly distinguishing between the two forms in his seminal paper of 1828, 'Mémoire sur l'influence que deux couleurs peuvent avoir l'une sur l'autre quand on les voit simultanément'.[136] Here and in his *magnum opus*, *De la loi du contraste simultané des couleurs* of 1839, Chevreul also made it clear he believed that it was 'simultaneous contrast' alone that generated coloured shadows.[137]

Although Chevreul did accept that coloured shadows required two light sources, he argued – like Field – that the coloured light source induced its complementary colour in the shadow lit by the neutral source and simply ignored any effect its surround may have had.[138] In the 1839 text, for example, he described how a white plaster figure illuminated by 'diffuse daylight' and 'coloured rays' will appear 'complementary in colour to the coloured light' in its shadows.[139] Similarly, Chevreul argued that the 'blue' colour 'shadows' take on when the sun is low and produce solely the sun's 'orangey' light.[140] The principle at issue, he added, was proven by the fact that 'red, yellow, green, purple' coloured light generated correspondingly 'green, purple, red, yellow' shadows.[141]

It is often suggested that Delacroix was influenced by Chevreul's ideas, and he is known to have borrowed some notes taken from the lectures the chemist gave at the Gobelins in 1848.[142] Delacroix may have taken an interest, too, in lectures Chevreul gave in the early 1830s, which were published in the *Magazin Pittoresque* in 1834.[143] The painter never met the scientist, however, having been prevented from attending the only rendezvous they arranged, around 1850, by his chronic sore throat.[144]

Alexandre Dumas even suggested in a *causerie* published in 1864 that Delacroix had 'described the law of the simultaneous contrast of colours' as early as 1827, 'several years before M. Chevreul'.[145] According to Dumas, Delacroix made his discovery when he called a cab in order to study the Rubenses in the Louvre for a solution to a problem he was having with the gold 'mantels' of the ancillary figures in *The Execution of the Doge Marino Faliero*, which was in fact painted between 1825 and 1826 (Figure 4.14).[146] 'Delacroix stopped short in front of the door of his cabriolet', Dumas

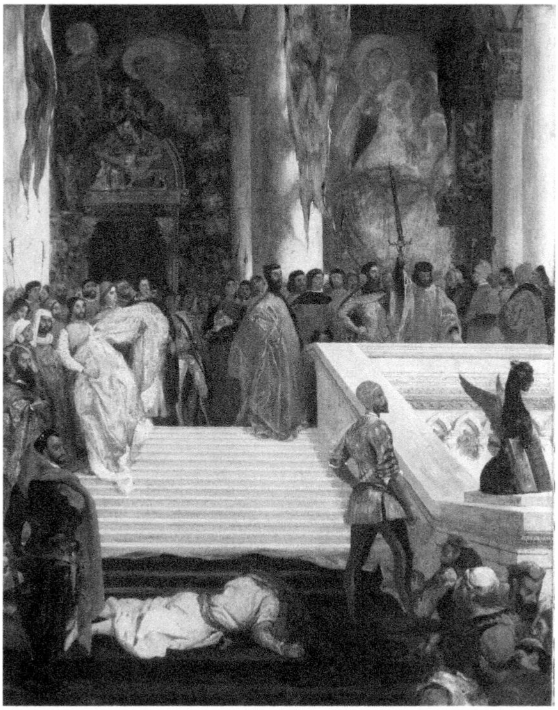

Figure 4.14 Eugène Delacroix, *The Execution of the Doge Marino Faliero*, 1825–26. Oil on canvas, 145.6 × 113.8 cm. London, Wallace Collection, inv. no. P282.

Source: Wallace Collection/Bridgeman Images.

recounted, not only because its colour was the 'most ferocious yellow imaginable', but also because the 'violet shadows' it cast made this 'brilliant' colour 'stand out' even more.[147] Strictly speaking, this implies Delacroix believed that shadow colour was complementary to the local colour of the surface next to it. And if so, then his theory is not consistent with Chevreul's.[148]

Dumas recalled that Delacroix's experience caused him to install a 'double triangle on the wall' of his studio to specify the complementary colours.[149] And an example of such a diagram was found among Delacroix's papers (Figure 4.15).[150] There are also passages in *Marino Faliero* which suggest he employed this diagram to calculate the colour of shadows. More particularly, the green shadow induced by the red colour in the Doge's shot silk mantel and the robes of the figures next to him follow the scheme of contrasts it embodies. It should be noted, however, that the predominant golden colour of these robes does not induce any purple in their shadows.

Later, Delacroix wrote beneath a colour triangle which he drew in his Moroccan notebook of 1832 that 'red has green shadows', and he noted with regard to the 'heads of two small peasants' he had observed that 'The yellow one had violet shadows' and the other 'redder' head had 'green shadows.'[151] The principle implicit in this statement is that shadow colours are complementary to the (subtractive) primaries (red, yellow, and blue) normally reserved for local colours and hence must be secondary colours (or green, purple, or orange).[152] The problem with this rule is that it does not accommodate the possibility that an object whose colour is a secondary can cast a

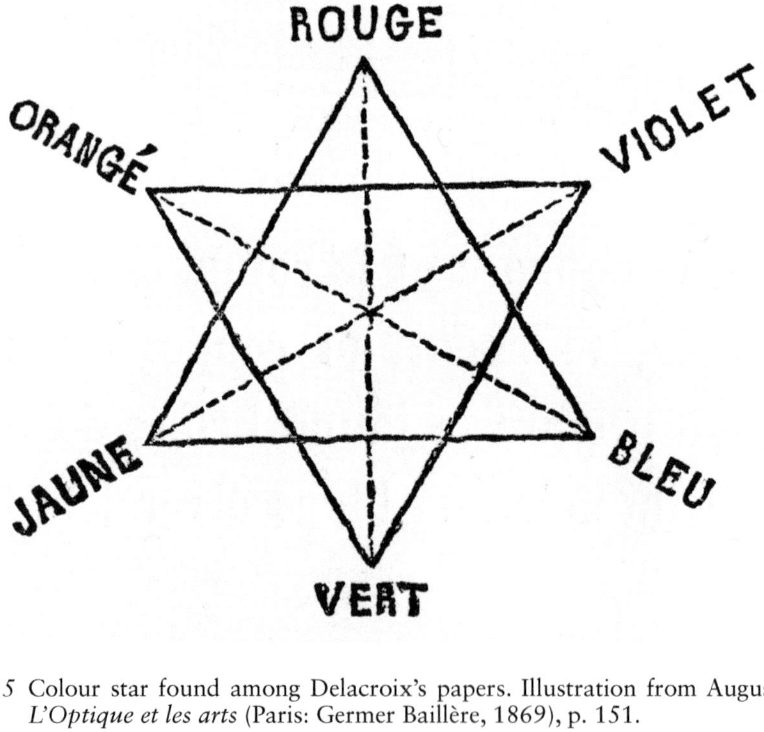

Figure 4.15 Colour star found among Delacroix's papers. Illustration from Auguste Laugel, *L'Optique et les arts* (Paris: Germer Baillère, 1869), p. 151.

Source: Author.

primary-coloured shadow. And significantly, Delacroix did not always follow it. In *Heliodorus Driven from the Temple* of 1854–61, for example, at the same time as the blue tunic of the figure in the bottom right corner exhibits orange shadows, the green cloak of the figure to his right displays red shadows.[153]

It should be emphasised that Delacroix developed a quite different theory of shadow colour in later life, in which contrast plays no role. Thus, according to the so-called law of green elaborated in the notes he made at Dieppe in 1854, 'Reflections are green' and 'shadows are violet' of their very nature.[154] Delacroix appears to have reiterated this argument in the same notebook when he recorded how, 'I see from my window the shadows of people passing by in the sunlight on the sand on the harbour. The sand on the ground here is violet of itself.'[155] He then mentioned how the colour of the shadow engendered its complementary in the sand, writing: 'The shadow of these characters is so violet that the ground becomes yellow'.[156] In this account, then, the already violet shadow is not influenced by its surround, but influences it.[157]

Although *De la loi du contraste simultané des couleurs* was not republished in French until 1889, it was a source of national pride, and its arguments were frequently referred to, even if sometimes inaccurately. Alfred de Lostalot probably had Chevreul in mind when he described how Monet made shadows complementary in colour to the light in a review from 1883. This stated that: 'The brilliance of the yellow rays exalts the nervous sensibility of the painter' and 'then produces in him the well-known physiological phenomenon wherein the complementary colour is evoked; so he sees violet'.[158] But although several witness accounts suggest that Monet was aware of ideas about contrast, despite claiming he had a 'horror or theories', there is no clear evidence that he read Chevreul's text.[159]

Pissarro did read Chevreul, however, as is attested by the letter of November 1886 to the critic, Félix Fénéon, in which he acknowledged that the Neo-Impressionist technique he had recently adopted was indebted to the 'theory of colour discovered by M. Chevreul'.[160] And in a letter to his son, Lucien, in February 1887, Pissarro linked his own understanding of light and contrast to Chevreul explicitly, declaring that:

> [I]f I did not know how colour behaved from the discoveries of Chevreul and other scientists, we [the Impressionists] would not have been able to pursue our study of light with so much confidence. I would not have drawn a distinction between local colour and the illumination, if science had not alerted us to it. The same goes for complementaries, and contrasts, etc.[161]

It is likely, then, that Chevreul's arguments stand behind the coloured shadows in paintings such as *Festival at l'Hermitage* of 1876–78 (Figure 4.16). Here, for instance, the shadow on the right-hand wall of the building in the middle distance is distinctly blue–violet and complementary to the orangey-yellowish light striking the adjoining wall.

A common point of reference for many painters interested in coloured shadows, including Van Gogh and Seurat, was the 'chromatic rose' or colour star in Charles Blancs's widely read manual, the *Grammaire des arts du dessin* of 1867 (Figure 4.17).[162] Blanc argued that this diagram illustrated the colours produced by simultaneous contrast, citing Chevreul to the effect that putting a 'colour on a canvas' is to 'colour the surrounding space with its complementary'.[163] Unfortunately, in the same passage, Blanc assimilated this effect to Goethe's description of the successive

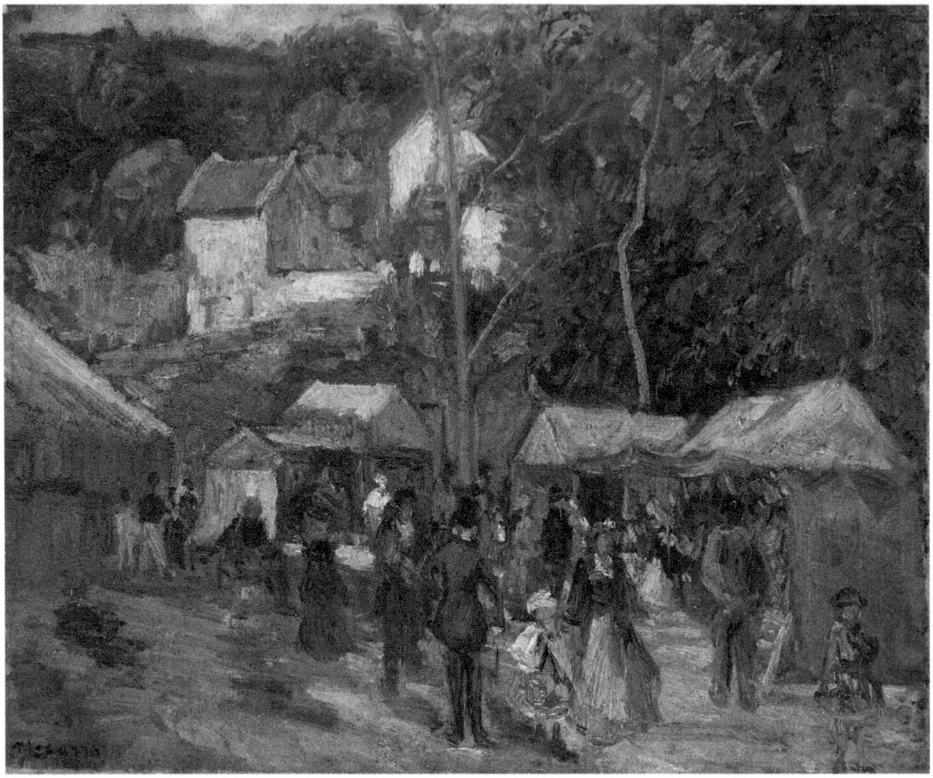

Figure 4.16 Camille Pissarro, *Festival at l'Hermitage*, 1876–78. Oil on canvas, 55.1 × 46.5 cm. London, The Courtauld Gallery, inv. no. P. 1978.PG.318.

Source: The Samuel Courtauld Trust, The Courtauld Gallery/Bridgeman Images.

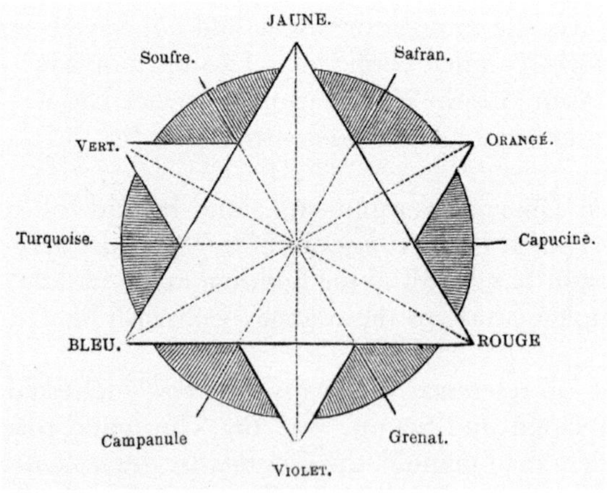

Figure 4.17 *Chromatic Rose*. Illustration from Charles Blanc, *Grammaire des arts du dessin* (Paris: Renouard, 1867), p. 600.

Source: Author.

contrast effect at work when 'yellow' crocuses engender 'violet' after-images.[164] And to make matters worse, he then likened both effects to a constancy effect Monge had reported when he described an experiment involving a green spot produced by red light transmitted through a curtain.[165] Blanc even described Delacroix's experience with the yellow cab as an example of simultaneous contrast, arguing that it (somehow) demonstrated the principle that 'shadows' are 'always lightly coloured by the complementary of the lights'.[166] In short, then, Blanc offered Chevreul's theory as an explanation of several distinct phenomena, including coloured shadows, which he mistakenly placed under the same rubric.

The American scientist and Sunday painter, Ogden Rood, who was an adherent of the additive theory of colour advanced by Maxwell and Helmholtz, subscribed to the theory that contrast generated shadow colour by colouring its surround (rather as espoused by Rumford and Goethe had argued). In his *Modern Chromatics* of 1879, for instance, he described a version of the familiar experiment with a candle and daylight, explaining that the shadow illuminated by white daylight – number 2 in the accompanying diagram (Figure 4.18) – appeared 'decidedly blue' because of its 'contrast with the surrounding orange-yellow ground' coloured by the candlelight.[167]

Although Rood specified that the process at work here was 'simultaneous contrast', he wrongly believed that the 'effects of simultaneous contrast' were 'identical in kind with those generated by successive contrast'.[168] He consequently claimed that the complementary colours generated by both varieties could be predicted by the contrast diagram he based on the laws of additive mixture.

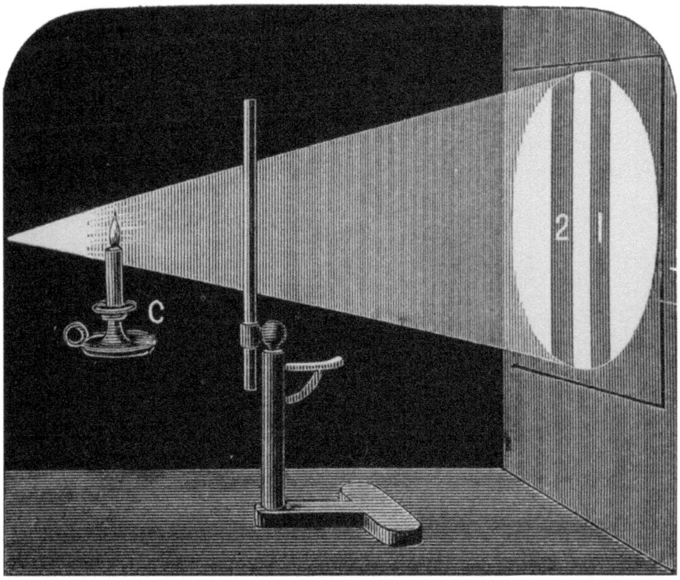

Figure 4.18 Shadows of Rod, Using Daylight and Candle-Light. Fig. 122 from Ogden N. Rood, *Modern Chromatics* (New York: Appleton, 1879), p. 255.

Source: Author.

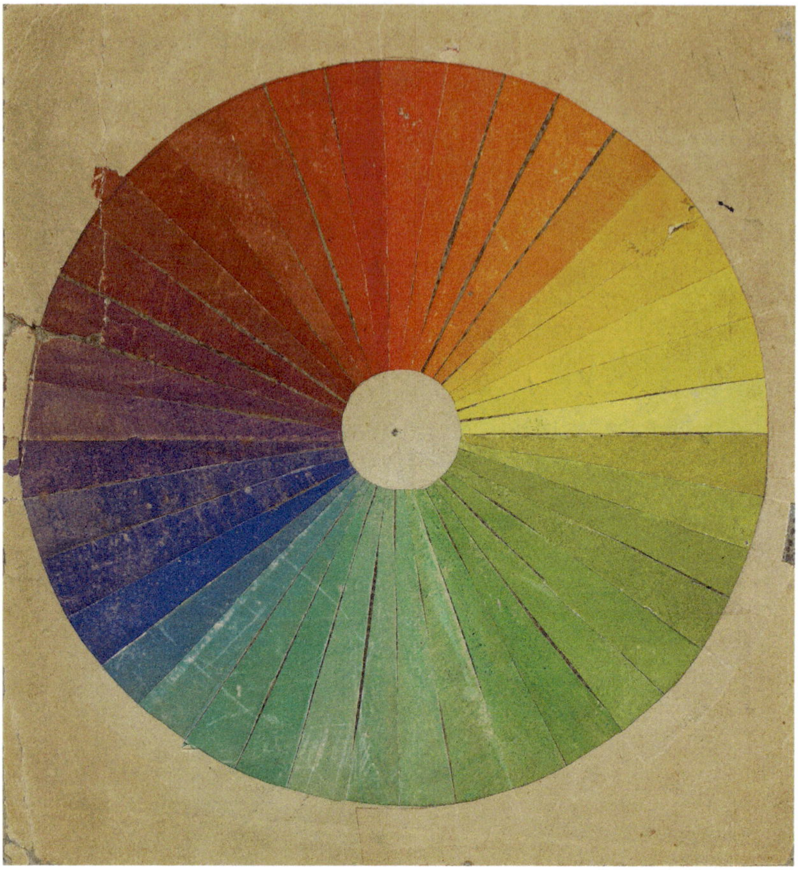

Figure 4.19 Louis Hayet, *Colour Wheel*, 1886. Watercolour and body colour on paper, mounted on dark grey board, 26 cm diameter. Oxford, Ashmolean Museum, acc. no. WA1979.17.

In December 1886, Louis Hayet sent Pissarro a colour circle based on Rood's diagram (Figure 4.19), now available in French, which used pieces of coloured paper to set out the colours the original represented in black and white.[169] In this additive scheme, the complementary to orange is cyan, a distinctly greenish blue. Hayet nevertheless ignored Rood's advice on his *Still Life with Oranges* of *c*. 1889 (Figure 4.20), since the orange light in this casts blue shadows. It would seem, then, that Hayet employed the subtractive logic in this work embodied in Blanc's chromatic rose, which he also copied.[170] It also appears that it is the light itself that creates the shadows' colour, as Chevreul claimed. If nothing else, then, these facts suggest there was a good deal of uncertainty among painters over the precise nature and specific causes of the colours that contrast induced in coloured shadows.

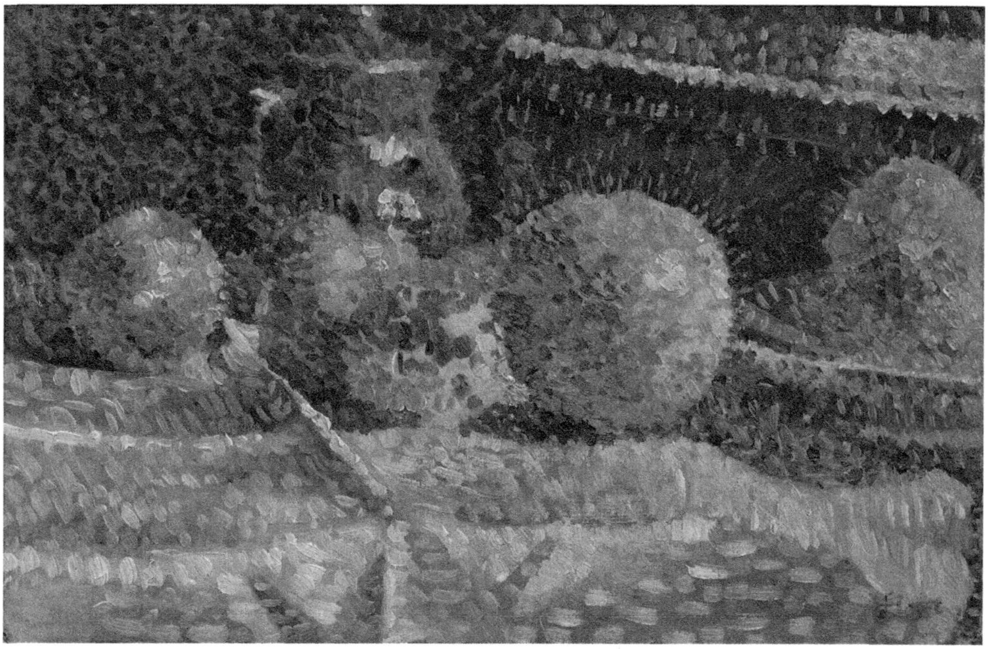

Figure 4.20 Louis Hayet, *Still Life with Oranges*, *c.* 1889. Oil on cardboard, 25 × 38 cm. Private collection.

Source: AKG Images.

Chevreul in England and America

Chevreul's memoir was translated into English in 1848, albeit in an anthology published by the Cavendish Society which is almost unknown today.[171] Similarly, although the two translations of *De la loi du contraste simultané des couleurs* by Charles Martel (the pseudonym of Thomas Delf) and John Spanton had run through at least 11 editions by 1890, the first (Delf's) did not appear until 1854.[172] So, although Hunt owned a copy of Chevreul's memoir, there is no evidence to suggest he knew about Chevreul's writings when he first painted coloured shadows around 1850.[173] Brown may have done so, however, since he spoke French.[174]

Hunt and his Pre-Raphaelite associates could have known Chevreul's ideas indirectly, nonetheless, from the section on 'Simultaneous Contrast' in Twining's *Philosophy of Art*, which in 1849 was the first English text on art to take the French scientist's work as its 'guide'.[175] Here, more specifically, Twining argued that contrast, which he misleadingly called 'comparison', could create or enhance shadow colour. 'In the effects of sunset', for instance, Twining maintained that the 'reflexes or secondary lights of the shadowy parts, appear cold, compared with the warmth of the sunshine'.[176] In the same vein, he argued that

> [i]f the prevailing tone of the lights be cold, the deepest shadows will appear warm, by comparison; and if the prevailing tints of the lighted portion be warm, the

local darks, if there be any reflexes, will appear, by comparison, of a somewhat colder hue.[177]

Whether or not Brown read Chevreul, he was an exponent of contrast. In *The Exhibition of Work*, for example, he described how *The Hayfield* of 1855 represented a scene in which 'rain' had so 'heightened the green' of the grass and the 'brown of the hay', that the 'hay by contrast with the green grass' appeared 'positively red or pink'.[178] And later, Ford Madox Hueffer claimed that Brown had depicted the effect of contrast on shadows more particularly, arguing that he 'first painted bright purple haycocks' upon a 'bright green field' because 'he happened to notice that when sunlight is rather red and the sky very blue, the shadowy side of green hay is all purple.'[179]

The fact that Brown 'improved' the 'Lady's mouth' in *Pretty Baa-Lambs*, by intensifying her red lips to make the shadowy teeth behind them look green, has also been linked to Chevreul.[180] In *Work* of 1852–65 (Figure 4.21), moreover, there are distinctly green shadows in the teeth and neck of the navvy leaning slightly backwards, which are evidently generated by his ruddy complexion. The equally saturated blue shadows in the lining of the navy in the centre are clearly the result of contrast as well. At least, they are not washed out or absorbed by its lemon–yellow surface, as they would be if they were simply physical in origin – as are those in the spade above his head, which mirrors the sky.

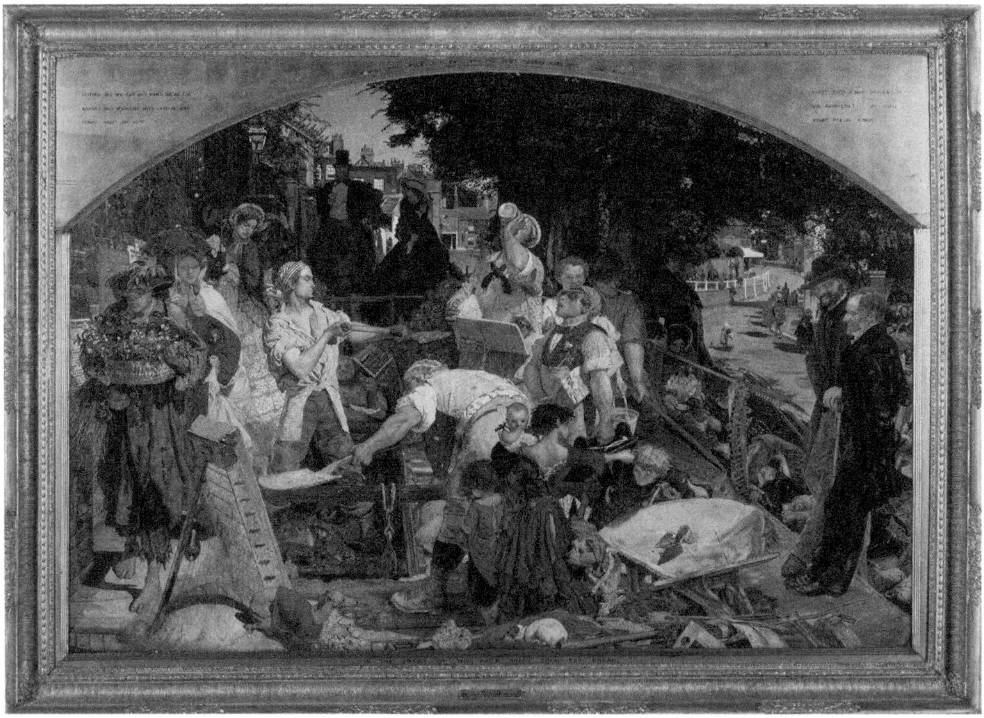

Figure 4.21 Ford Madox Brown, *Work*, 1852–65. Oil on canvas, 137 × 197.3 cm. Manchester, Manchester Art Gallery, acc. no. 1885.1.

Source: Manchester Art Gallery/Bridgeman Images.

Another source for Brown's understanding of coloured shadows may have been the popular book, *Colour as a Means of Art* of 1838, by Frank Howard, an artist with whom he shared studio accommodation in 1844 and whom he continued to see at least until 1848.[181] The fact that Howard dedicated the book to Callcott may even indicate that Rumford's ideas about such phenomena lay behind his understanding of them.

Howard's text is prolix and confused, but it does make a coherent case that 'contrast' produces coloured shadows when a 'yellow tinge in the lights' makes 'shadows appear to incline to purple'.[182] And, in a section devoted to the 'COLOURS OF LIGHTS AND SHADOWS', Howard argued clearly enough that 'From the effect of contrast, shadows appear *comparatively* of the opposite colour to that of the light'.[183] Notwithstanding, he then described how the 'Bianchi' (painters such as 'Rubens, Vandyke, and Lawrence' who aim to imitate the intensity of natural light) depicted shadows in the colour complementary to that of the surface on which they fell, rather than the light, rendering 'shadows on green objects in the foreground', for example, in 'dark crimson'.[184] Howard nevertheless included a plate, *Sunshine*, in which the foreground shadows have this colour or perhaps the 'purply brown, broken with red reflections' the Bianchi also favoured[185] (Figure 4.22).[186]

By the 1860s, the idea that contrast generated complementary coloured shadows had become a cliché. The watercolourist John Chase, for example, wrote in his

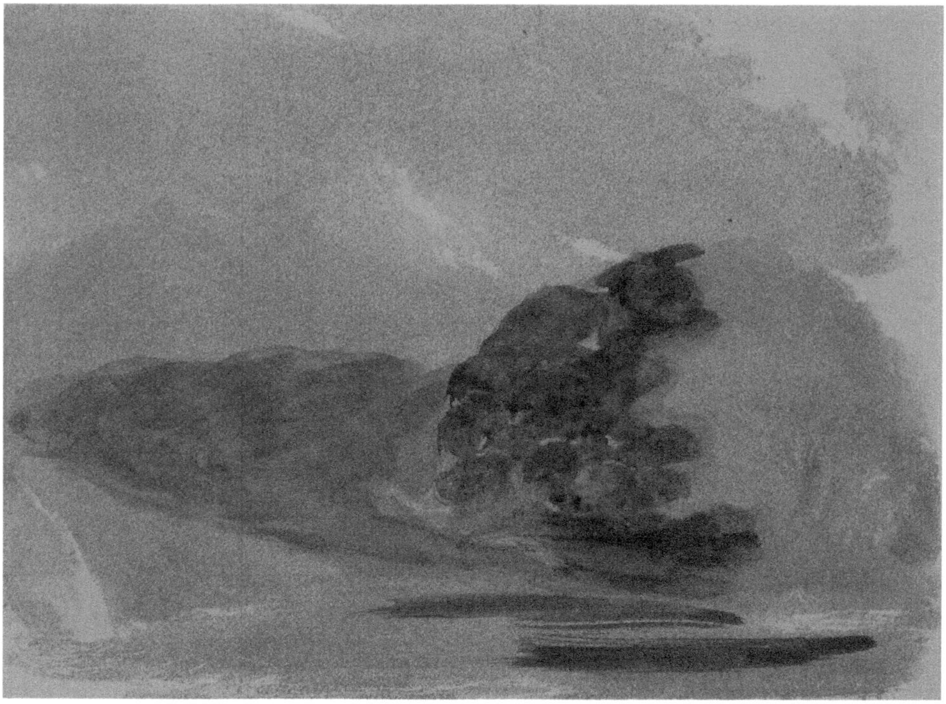

Figure 4.22 *Sunshine*. Plate 15 from Frank Howard, *Colour, as a Means of Art* (London: Joseph Thomas, 1838), opposite p. 99.

Source: Author.

manual, *A Practical Treatise on Landscape Painting* of 1861, that 'In sunset scenes all parts in light are tinted with orange; the complementary, or harmoniously contrasting colour to orange is blue; we therefore find in nature shadows of bluish grey contrasting with orange light'.[187] And he illustrated the effect with a chromolithographed plate, *At Chevening, Kent*. It would seem, therefore, that even if Chevreul's ideas lay somewhere behind this statement, anonymity was the price he paid for their success.

The American painter, John La Farge, was both an admirer of the Pre-Raphaelites and a devotee of Chevreul's.[188] He bought a copy of one of the English translations of Chevreul while in London in 1857.[189] And later, he recalled that what he 'read in the writings of the illustrious Chevreul' about 'optical views on color' determined 'more than anything else the direction' his 'painting took some years afterward'.[190] In addition, La Farge told Celia Waern that:

> My youthful intolerance required the relations of color for shadow and for light to be based on some scheme of color-light that should allow oppositions and gradations representing the effects of the different directions and intensities of light in nature, and I already became much interested in the question of the effect of the complementary colors.[191]

La Farge made several other statements about his interests in colour science.[192] It seems likely, then, that Chevreul's ideas did, as some have suggested, play a role in defining the colours of *The Last Valley – Paradise Rocks* of 1867–68 (Figure 4.23).[193]

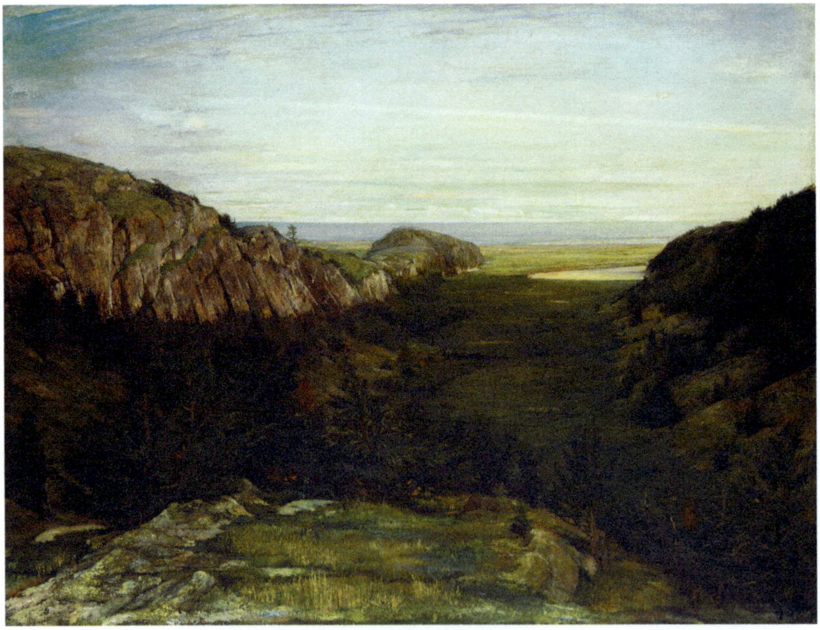

Figure 4.23 John La Farge, *The Last Valley – Paradise Rocks*, 1867–68. Oil on canvas, 82.7 × 103.4 cm. Washington, National Gallery of Art, Gaillard F. Ravenel and Frances P. Smyth-Ravenel Fund, acc. no. 2000.144.1.

Source: NGA Images.

Other remarks of La Farge's suggest, however, that his paintings had only a loose relationship to theory. He told Waern, for instance, that the 'underpainting of shadows' was 'blue' and 'that of the lights red', a pairing which is not complementary in Chevreul's scheme.[194] Moreover, when La Farge gave a series of lectures on painting in 1893, the only reference he made to a coloured shadow quoted the physicalist description that Eugène Fromentin had given, in *Une année dans le Sahel* of 1859, of a 'shadow, inundated by reflections from the sky'.[195]

Winslow Homer met La Farge in the second half of the 1860s, and the two often encountered each other from 1872, when Homer moved into the studio building on Tenth Street his colleague had frequented in the 1860s.[196] In his obituary for Homer, La Farge also recalled that at the time of their association:

> I was just beginning to study in the direction of the future question of coloured light and the relations of complementaries. Of course, a great deal had been done that way, but not as yet as far as we proposed to carry it.[197]

Homer shared La Farge's admiration for Chevreul in particular, since he owned two copies of the Spanton translation. According to an inscription, he was given his first copy (of the 1859 edition) in July 1860 by his brother, Charles, who had purchased it in London.[198] And in 1903, Homer told John Beatty that this copy was his 'Bible'.[199] The still extant copy bears out this statement since, just as Beatty recalled, it is 'well worn and annotated', Homer having made many additions to it between 1860 and 1884.[200] Lloyd Goodrich also recounts an anecdote which implies that the ageing Homer applied Chevreul's theories in the garden of his studio at Prout's Neck in Maine. This relates how a

> young neighbor working in his flower garden was once accosted by the painter with the query, 'Who taught you about color?', and on his next visit to the studio Chevreul was produced to show that he had arranged his flowers exactly right.[201]

Although Homer's annotations do not relate to the single passage on coloured shadows in Spanton's edition, it is possible that Chevreul's influence manifested itself in the vivid blue shadows of works visible in many of Homer's works, including *Farmyard Scene* of *c.* 1872–74 (Figure 4.24). At least, judging from the colours of the foremost gable wall, it could be that Homer used a reddish-orange to represent the action of the setting sun and blue to depict the near-complementary shadow colour this would have induced. It is more likely, however, that homer simply regarded shadows as coloured by the light of the sky, like La Farge. The fact that a painter read Chevreul is in any event no proof that he either understood, or implemented, the chemist's ideas.

A similarly idiosyncratic take on Chevreul's ideas is evident in the book, *Color Problems* of 1901, by the New York painter, Emily Noyes Vanderpoel. This discussed the chemist's ideas at length, particularly in sections on 'Contrasts and Complements' and 'Color-Harmonies'.[202] In the first of these, Vanderpoel claimed – *contra* Chevreul – that the 'complementary of a color exists in its shadow' and that this principle is demonstrated by the fact that a 'shadow' in 'snow', which is 'grey' when the 'sun is hidden by a cloud', turns 'more or less blue' when the cloud 'passes' and the sun's 'light on the snow makes it look yellow'.[203] Vanderpoel nevertheless insisted that the 'purple or violet shadows of the "impressionists"' were 'in many cases exaggerations', not least – she argued – because they failed to accord due

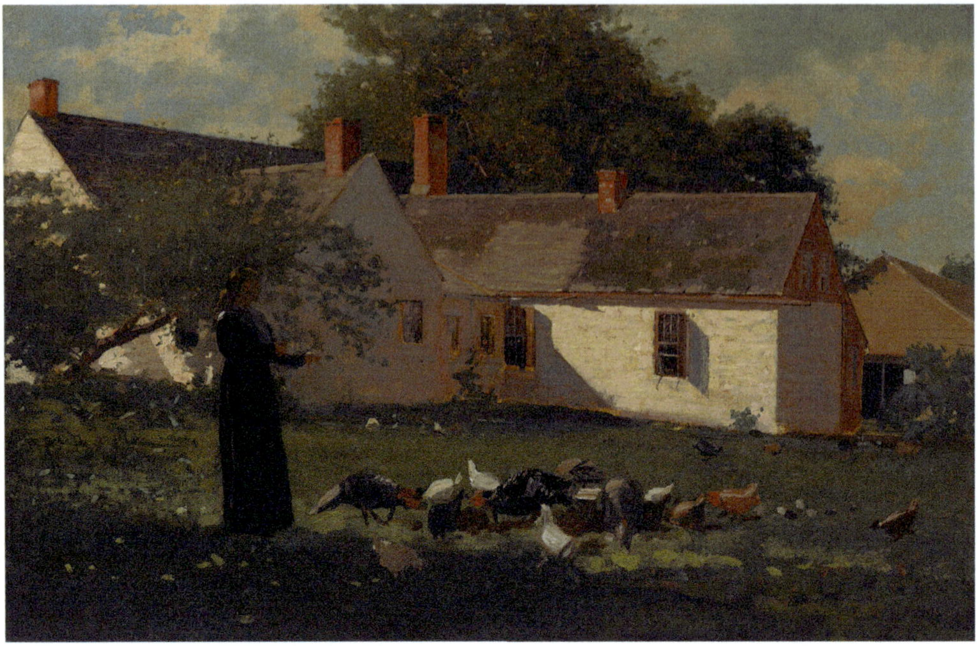

Figure 4.24 Winslow Homer, *Farmyard Scene*, *c.* 1872–74. Oil on canvas, 31.4 × 46.8 cm. Williamstown, Sterling and Francine Clark Art Institute, inv. no. 1955.772.

Source: Clark Art Institute/Bridgeman Images.

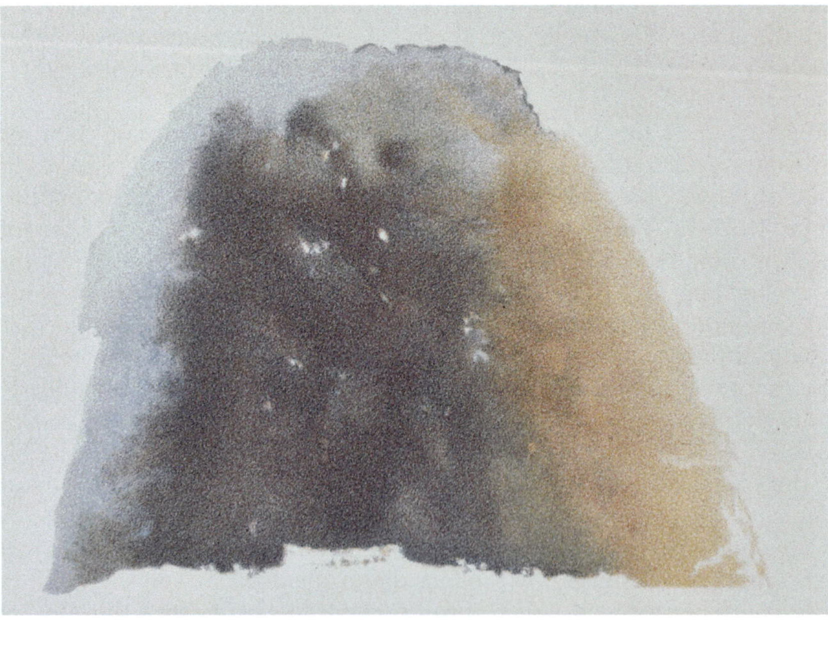

Figure 4.25 *Color Note from a Shadow on White Ground*. Plate CV from Emily Noyes Vanderpoel, *Color Problems*. 3rd edn (New York, London, Bombay: Longmans, Green & Co., 1903).

Source: Author.

weight to the 'color of the sky reflected in the shadow'.[204] Vanderpoel even added an intriguing colour plate to her book (Figure 4.25) illustrating a coloured shadow.[205] But this was too abstract to convey her ideas about the conditions that actually created such effects.

Reflection and Contrast

It is clear from their statements and their works that several painters – Valenciennes, Brown, and Hunt among them – believed that coloured shadows could be produced both by reflection and by contrast. Even Goethe, who abandoned physicalism for the theory of contrast in the *Farbenlehre*, never completely relinquished his earlier 'extreme view', according to Eastlake.[206] Citing a conversation recorded by Johann Peter Eckermann in February 1831, he noted how Goethe in fact conceded that 'in the case of the blue shadows of snow, "Both causes may, however, co-operate"'.[207]

Statements like this can considered coherent if taken to imply that two different causes can produce the same phenomenon, either separately or in concert. They nonetheless fail to accommodate how in fact physical factors and effects of contrast are only ancillary to the cause which is primarily responsible for creating coloured shadows, namely constancy. To develop and hold a theory capable of expressing this fact, however, required embracing a much greater degree of complexity than either the theory of contrast or of reflection involved. So although neither theory could properly explain coloured shadows, the very simplicity of the explanations they did provide may account for why they held sway among artists for such a long time and why they clouded the judgment of so many distinguished scientists.

Notes

1. See Baxandall 1995, 78 and 115.
2. Jombert 1755, 103–104.
3. Harris 1770, 8.
4. Ibid.
5. See Koenderink et al. 2020.
6. See Harris 1770, 4.
7. Ibid., 8. On Newton's colour circle, see Newton 1704, 114–117.
8. Jombert 1755, 104.
9. Valenciennes 1799, 289–290. Cited in Dorra 1994, 189. Valenciennes's explanation also implies a physical explanation of the phenomenon, since a cloudy sky reflects less blue light than a clear one.
10. Valenciennes 1799, 442.
11. Darwin 1791, 1: 8 (of the 'additional notes' following 214).
12. Thompson 1794, 107. This paper is quoted at length in Cohen 2001, 112–112.
13. Ibid., 107–108.
14. Ibid., 109.
15. Ibid., 115.
16. Ibid.
17. Ibid., 116.
18. Ibid.
19. Ibid.
20. Thompson 1802, 334.
21. Ibid.
22. Thompson 1802, 335. On the notion of complementarity elaborated in this paper, see Finley 1967, 360 and Cohen 2001, 113–114.

23. Thompson 1802, 339.
24. Ibid., 115.
25. Ibid., 114–116.
26. Thompson 1799.
27. Callcott 1810–20, f36r–37v. On the group of manuscripts to which this belongs, now in the Bodleian Library but then in the Ashmolean Museum, see Gage 1993, 109, 204, and 283 n. 111 and Gage 2001, 21, 62, 66 n. 78, 69 n. 146. I am grateful to Colin Harrison for his help with locating these.
28. See Field 1811 and Field 1817. On the manuscript (Ms 8), see Gage 1989, 30, 40 and 81 and Gage 2001, 9, 62, 66 n. 72, 69 n. 149, and 70 n. 171.
29. Field 1811, f58r; citing Galton 1799, 513. See also Field 1811, f57v (the page opposite the Galton quotation), which cites the version of Rumford's essay published in his *Philosophical Papers* thus: 'See Rumford's Ph. Essays, V. I, P. 319. 1802'. All extracts from Field's notebooks are by the kind permission of Winsor & Newton.
30. Field 1811, f58r.
31. Thompson 1797.
32. Ibid., f74r.
33. Field 1811, f74r.
34. See Goethe 1810, 2: 575 and Sepper 1988, 89 (for the remark Goethe made to Lichtenberg to this effect).
35. Joost et al. 2002, 302.
36. Ibid., 303.
37. See Goethe 1840, 29–30.
38. Ibid., 30.
39. Ibid., 32.
40. Ibid., 30.
41. This study is in the collection of the Klassik Stiftung, Weimar. Inventory number: GFz 127, *Zur Farbenlehre, Tafel I, Abwandlung des dritten Entwurfes*, 1806–10.
42. Goethe 1840, 34. See also Ott 1977, 11–12, for a criticism Eckermann made in 1829 of Goethe's claims about the 'subjective' nature of coloured shadows in snow.
43. Goethe 1840, 34.
44. Ibid., 35.
45. Ibid., 36; paraphrasing Newton 1704, 139, which explained this phenomenon as follows: 'Sea-Water reflects back the violet and blue-making Rays most easily, and lets the red-making Rays pass most freely and copiously to great Depths', where 'the blue-making, green-making, and yellow-making Rays being reflected from below more copiously than the red-making ones, must compound a green'. See also Duck 1988, 515.
46. Goethe 1840, 36.
47. Ibid., 20–29.
48. On augmentation, see ibid., 99, 212, and 214; Gray 1952, 114; Matthaei 1971, 37; and Burwick 2012, 19.
49. Goethe 1840, 35.
50. Ibid.
51. Ibid., 36.
52. Ibid., 21.
53. Ibid. Eastlake's 'demands' might be better put by 'calls for'; but I have used his familiar term throughout. See also ibid. 39, for the claim that 'Candle-light at twilight acts powerfully as a yellow light: this is best proved by the purple blue shadows which, under these circumstances, are evoked by the eye'.
54. Ibid., 28. See Schöller 2017, 20.
55. On this issue, see Tantillo 2002, 37–47 and Muenzer 2007, 222–225. See also Gray 1952, 106.
56. Runge 1840, 1: 169.
57. See ibid., 2: 318–320 for a letter to Goethe of September 1806 mentioning Newton.
58. For the first and last letters in this sequence, see Goethe 1887–1912, 19: 131–132 and 21: 215–216 and Runge 1840, 1: 308–309 and 2: 409.

    Runge responded to Goethe's invitation the same month by sending him a long letter on colour theory, which was subsequently included in the *Farbenlehre*. For the letter, which also contained some remarks about transparent colours which Wittgenstein later took up, see ibid., 1: 88–98 and Goethe 1810, 1: 339–349. On its inclusion in the *Farbenlehre*, see

Goethe 1887–1912, 19: 451 and 21: 188 and Runge 1840, 2: 351 and 388. For letters from Goethe to Runge not cited, see Goethe 1887–1912, 19: 178–179 and 230; 20: 119 and 204–205 and Runge 1840, 2: 308–309, 315–316, 329, 363, 370, 388–389. For Runge's letters to Goethe, see ibid., 1: 177 and 2: 310. On the relationship between the painter and the poet, see Mitchell 59 and Allert 2007.

59. See ibid., 1: 155 and 180.
60. See ibid., 1: 184 and 200. For Runge's desire to see Goethe's 'theory of colours' or 'treatise on colour', see ibid., 2: 331 and 350. See also ibid., 1: 155–156 and 159–160, and 2: 344 and 2: 350 (for a letter in which Runge mentions a letter from Goethe concerning the *Farbenlehre* which is not extant).
61. For Runge's remarks on 'contrast' not cited here, see Runge 1810, 22 and Runge 1840, 1: 58, 127.
62. Ibid., 2: 330–331.
63. Ibid., 1: 169.
64. Ibid., 1: 154.
65. Runge 1810, 26.
66. Ibid.
67. See Goethe 1840, 1, 28, 35, 225, 280, 317, 319, and 343.
68. See Börsch-Supan 1974, 61, 63, and 79.
69. See Carus 2002, 8. On the relationship between Carus and Goethe, see Allert 2012, 45–46 n. 25 and 47 n. 39; and Allert 2016.
70. See Carus 2002, 7–88. For Goethe's letters to Carus, see Goethe 1887–1912, 29: 96–100 and 321; 33: 87–89; 35: 233–234 and 238; 36: 22–23, 27–28, 54, and 292–293; 37: 12–13, 16–17, 21, 32, 141, and 248–249: 38: 2–4 and 259–261; 39: 88; and 40: 228.
71. Goethe 1887–1912, 36: 22–23. A second edition, incorporating one additional letter, was published in 1835. See Carus 2002, 46.
72. Ibid., 85; and Goethe 1840, 2–59.
73. Carus 2002, 141.
74. See Richter 2016, 40; and Kittelmann 2016, 198.
75. On Kopisch's interest in light and colour, see Kittelmann 2016.
76. Goethe 1840, 36.
77. Goethe 1887–1912, 25: 180–182. The unpublished manuscript of a letter from Roux to Goethe of 1821 is in the Goethe and Schiller Archives in the Klassik Stiftung, Weimar. Inventory number: GSA 28/94; Bl. 262–263.
78. For Goethe's letter about the art school, see Goethe 1887–1912, 31: 94. On Roux's career, see Heinstein 2005.
79. Roux 1824–29, 3: 76–77.
80. Ibid., 3: 77.
81. Ibid. See also ibid., 3: 79.
82. Ibid., 3: 74.
83. Ibid., 3: 78.
84. Ibid.
85. Ibid., 3: 79–80.
86. Ibid., 3: 79.
87. Ibid., 3: 95–97.
88. Schreiber 1868, 50.
89. Ibid.
90. Schreiber 1868, 51.
91. See Rovi 1980, 373. Bossi read the paper at the Instituto di Scienze et Lettere ed Arte of Milan.
92. Bossi 1821, 296.
93. Ibid., 297.
94. Ibid.
95. Ibid.
96. Monge 1789, 133–134. Monge's experiment is mentioned in Venturi 1801, 131 (one of Bossi's sources).
97. Bossi 1821, 298–299.
98. Ibid., 298. See Petrini 1815, 54, for a version of this experiment, in which objects viewed in the light 'transmitted' through a 'crystal' appear coloured by it, while those seen under the

light 'reflected' by it appear 'complementary' in colour. Although Petrini did not acknowledge Bossi, it is likely that he knew the artist's paper, since he also used the expression 'consequent colours' several times in the same book. See ibid., 29, 36, 38, 49, 56.

99. See also Bossi 1821, 292 for Bossi's claim that claimed that because 'shadows' are 'opposite' in colour to the real colour of surfaces, they are 'different' in colour for each of the coloured earths.
100. Ibid., 295. The first three are also mentioned in Venturi 1801, 103.
101. Ibid., 295. See also Rovi 1980, 376 n. 18.
102. Bossi 1821, 300. See also Goethe 1840, 21 and 28. For a similar argument, see Venturi 1801, 106.
103. Bossi 1821, 300. See also Rovi 1980, 380.
104. Bossi 1821, 304. On Bossi's other debts to Goethe, see Rovi 382. See also Bossi 1821, 295 for a description of experiments with 'coloured circles' against a white or grey ground which replicate those described in Goethe 1840, 11 and Bossi 1821, 304–305 on portraits painted in colours 'colours contrasting' to those of reality as described in Goethe 1840, 22–23.
105. Bossi 1821, 300, citing Goethe on 'pathological' colours. See Goethe 1840, 45–55. For the theory and the claim that these movements corresponded to the positions of the different colours on the colour circle, see Bossi 1821, 302–303. See also Rovi 1980, 380–382.
106. Venturi 1801, 105. The Italian reads 'tinte immaginarie consecutive' not 'tinte immaginarie conseguenti' as stated in Rovi 1980, 380. See also Venturi 1801, 111 for the scientist's ideas about 'the sentient fibre'; and ibid., 103–111 for his theory of the after-images produced by successive contrast. On Bossi's debt to this text, see Rovi 1980, 379–380. Most copies of Venturi's book belong to the 'edizione secunda' of 1801; but a few copies exist of an edition published in 1800 with no publisher's name or place of publication.
107. Bossi 1821, 303. See also ibid., 303–304.
108. Ibid., 316.
109. Calvi 1842, 67–68.
110. Ibid., 32–34.
111. For this claim, see ibid., 34–35. For the discussion, see ibid., 31 and 34–40.
112. Ibid., 29–30 and 50–57.
113. Ibid., 58.
114. Ibid.
115. Ibid., 59–61.
116. Field 1811, f72r.
117. Field 1835, 239.
118. Ibid.
119. See Bourgeois 1828, 179, which describes how introducing 'cylinder' into the beam of 'direct' (white) light overlapping the 'solar spectrum' produced by a 'prism' created 'two parallel bands of coloured shadows', one of which allowed the colours of the spectrum to show 'freely' (or without being desaturated by the white light) and another in which 'the exact complementaries of the hues in this first shadow' manifested themselves 'very clearly'.
120. Field 1811, f73r; citing Young 1807, 2: 787 (a description of 'The colours of a mixed plate', illustrated in Plate XXX, Fig. 450).
121. See Young 1807, 1: 470, which describes how viewing the edge of a dark object against a light background through a lens and a glass sheet generates two patterns of undulations in the light transmitted, which partially cancel each other out.
122. Field 1811, f73r.
123. Field 1845, 210–211.
124. Field 1835, 239 and Field 1845, 210. This passage, like some others in the first (1835) edition of the *Chromatography*, was omitted from the second (1841) edition of the book and was moved to the second (1845) edition of the *Chromatics* instead.
125. Field 1835, 239 and Field 1845, 210.
126. Field 1835, 239 and Field 1845, 210–11.
127. Field 1835, 132 and Field 1841, 241.

128. See Field 1835, v. Clover is mentioned in Field 1811, f94r and Field 1835, 259. See Gage 2001, 19–20 and 69 n. 150.
129. f25r within the sketchbook at NWHCM: 1939.141. On this sketchbook, see Gage 1989, 61 and Gage 2001, 19 and 69 n. 151.
130. See Gage 1989, 77 n. 33. See also Gage 2001, 14 and note 114. The same letter records that, in 1856, Hunt had bought a copy of the abridged version of this text, the *Rudiments of the Painter's Art*. See Field 1850.
131. See Bronkhurst 2006, 1: 156–157, which describes how Hunt scraped down some 'specks' and replaced them with 'stippling' upon his return to London. See also Staley 1973, 62–63, which states that Hunt worked on the painting until December, before returning to London. Delacroix was 'filled with wonder at Hunt's sheep' when he saw it at the Exposition Universelle. See Delacroix 1893, 3: 51 (journal entry for 30 June 1855).
132. See Field 1835, 32, and Field 1841, 61, for the claim that 'the colour of light is always opposite to that of the sky and shade', so that when 'the general colour of the sky is blue' the 'white which is to represent light should be tinged' with 'orange'. See also Field 1835, 239, and Field 1845, 211, for the more implausible claim that if 'when the sky is bluest, the shadows of an object be projected on white paper at a window opposite the north, into which the sun never enters, such shadows will be so far from blue, that they will be of a colour more or less warm, in proportion to the blueness of the sky'.
133. Field 1841, 40.
134. Prieur 1805, 7–8.
135. Ibid., 8.
136. See Chevreul 1832, 472–485, and 509.
137. See Chevreul 1832, 30–32 and Chevreul 1839, 60–63. On Chevreul's theory of coloured shadows as elaborated in the later text, see Schöller 2017, 29–34.
138. See Roque 2009, 247, which draws an analogy between Chevreul's theory and the idea advanced by Jean Henri Hassenfratz in his 'Premier mémoire sur les ombres colorées' of 1802 that the light sources producing coloured shadows are sometimes coloured reflections. For this proposal, see Hassenfratz 1802, 277.
139. Chevreul 1839, 447. See also ibid., 449 which described how a white plaster figure lit by 'diffuse daylight' from a window opposite and the 'coloured light' passing through a curtain placed over another window will appear 'white' in some places and 'tinted' with the colour 'complementary' to the filtered light in others. See also Lanthony 2006, 23–25, which summarises and illustrates Chevreul's argument.
140. Chevreul 1839, 680–681.
141. Ibid.
142. See Roque 2009, 241.
143. Ibid., 237
144. Signac 1899, 42. On Delacroix's implementation of Chevreul's principles, see Johnson 1963, 63–72.
145. Trimm 1864, 1. Dumas's lecture was delivered in the exhibition of Delacroix's work at the Société Nationale des Beaux-Arts. See Escholier 1926, 1: 66.
146. Trimm 1864, 1.
147. Ibid.
148. See Roque 2009, 247–248, for the opposite view. For an account of Delacroix's use of contrast in shadows and his putative reliance on Chevreul's principles, see Schöller 2017, 55–65. See Chevreul 1839, 236, for the argument that local colours can induce their complementaries in shadows in a painting, for example, when 'pinkish flesh colours' in portraits produce a 'greenish hue' in the 'shadowed parts' without any 'green' paint being used. See also Twining 1849, 239, on whether or not the 'mutual influence' of colours in the painting 'will exactly correspond with that they exercise over each other in the real subject'.
149. Trimm 1864, 1.
150. See Laugel 1869, 151. See also Gage 1993. 174; Roque 1994, 424 n. 60; and Roque 2009, 569 n. 13 and n. 19. On other colour diagrams of Delacroix's, see Silvestre 1864, 17; Johnson 1963, 56; Kemp 1990, 309; and Roque 2009, 239.
151. Cited in Roque 2009, 238 and 247. On Delacroix's diagram, see Johnson 1963, 64; Kemp 1990, 308; Roque 1994, 424 n. 59; and Roque 2009, 569 n. 15.

152. See ibid., 244–247.
153. See Johnson 1963, 104.
154. Piron 1865, 416. Cited in Gage 1993, 175. See also Roque 2009, 245–247. For similar remarks, see Delacroix 1893, 2: 142 (journal entry for 15 January 1853).
155. Piron 1865, 417–418. Cited in Johnson 2001, 226.
156. Piron 1865, 418.
157. See Johnson 2001, 226, which cites the argument offered in Roque 2009, 249, that – unknown to Delacroix – it was actually the yellow of the sand that induced the violet colour in the shadows.
158. Lostalot 1883, 346. Cited in Reutersvärd 1950, 109; Roque 1996, 36; and Roque 2009, 309. See also Kemp 1990, 311, for a statement on 'violet' shadows by Manet (although attributed here to Monet), citing Claretie 1881, 226.
159. Charteris 1927, 131; quoting a letter from Monet of June 1926. Cited in Rewald 1973, 586 and Roque 2009, 297. See also Roque 2009, 296–298, for accounts by Lostalot, Georges Lecomte, and Émile Bernard of Monet's familiarity with theories of contrast.
160. Pissarro 1980–91, 2: 75 [no. 358]. Cited in Roque 1996, 28–29 and Roque 2009, 286–287. See also Smith 1997, 23.
161. Pissarro 1980–91, 2: 131 [no. 397]. Cited in Smith 1997, 25.
162. See Blanc 1867, after 598 and 599 for colour and monochrome versions of this diagram. Seurat declared his awareness of Blanc's writings (without mentioning the *Grammaire*) in a letter of 20 June 1890 to Fénéon. On this and other sources mentioning his interest in Blanc, see Homer 1978, 17–18 and 268–269 nn. 20 and 21 and Herbert et al. 1991, 431–431. See Van Gogh 2010 for a letter [no. 454] to Anthon van Rappard of between mid-August and early September 1884, in which he mentions buying a copy of Blanc's *Grammaire* in 1884. On Van Gogh's interest in coloured shadows, see Schöller 2017, 71–84.
163. Blanc 1867, 599; citing Chevreul 1839, 197–198.
164. Blanc 1867, 60; citing Goethe 1863, 2: 126.
165. Blanc 1867, 600; citing Monge 1820, 184–185.
166. Blanc 1867, 600.
167. Rood 1879, 255.
168. Ibid., 254 and 257.
169. On this diagram, which was donated to the Ashmolean Museum by John Rewald, the series of 'discs' to which it belongs, and Hayet's and Pissarro's correspondence about these, see Thorold 1980, 14–15; Callen 1982, 109; Pissarro 1980–91, 2: 77 [no. 360]; Ward 1996, 288 n. 18; and Dulon and Duvivier 1991, 60–63.
170. The star forms part of the same collection of objects. See Thorold 1980, 14.
171. The text was contained in Chevreul 1848, 167–222. The remainder of the text was a translation of lectures Chevreul gave in Lyon in 1842–43.
172. Delf's was a complete translation of Chevreul's text. See Chevreul 1854. Spanton's contained only 636 of the original 1010 sections and omitted the second pair of sections in the original on coloured shadows (sections 680–681 in Chevreul 1839, 962). See Chevreul 1857.
173. I was alerted to Hunt's copy in the 1990s by my then colleague, Ed Lilley of Bristol University, who saw it in a bookshop in Crewkerne. See also Smith 2012, 20, which describes Hunt's depiction of 'simultaneous contrast' in *Our English Coasts*.
174. On Brown's fluency in French, see Bendiner 1998, 17, n. 30.
175. Twining 1849, 236–40. See Gage 2001, 72 n. 212 on Twining and Chevreul. See also Smith 2012, 237 n. 12, for the claim that Chevreul's 1839 text was 'first' discussed in Sweetlove 1852, 7, which in fact conflated simultaneous and successive contrast, while taking Chevreul to task for its own confusion.
176. Twining 1849, 195.
177. Ibid.
178. Brown 1865, 10. Cited in Parris 1984, 133–144 and Staley 1973, 40 (in part). See also Bennet 2010, 1: 182 and Treuherz et al. 2011, 170, on how his dealer, White, had refused to buy it on account of its 'pink' hay.
179. See also Hueffer 1911, 213. Cited in Thirlwell 2009, 33.

180. See Parris 1984, 94; Townsend et al. 2004, 130; Bennett 2010, 123, and Smith 2012, 20, which argues that Brown represented the 'after-image' of her red 'lips and cheeks', indicating his 'intuitive awareness' of the 'laws of simultaneous [sic] contrast' described by Chevreul. Smith also makes the observation that the effect 'is only noticeable on close inspection: from a distance the effect appears natural and correct'. For the evidence that we are more likely to notice colour when it occupies a substantial portion of the visual field because we perceive it at low spatial frequency, see Livingstone 2002, 194–195; and Mather 2013, 22, 118, and 120.
181. For Brown's occupancy of a studio in Tudor Lodge (in Arlington Street, Camden Town) in 1844, see Hueffer 1896, 37–38, which also records that Howard 'had the largest' studio in this 'nest'. See also Rossetti 1900, 52, for a letter from Brown of December 1844 concerning the 'infernal' but 'good-natured' Howard; and 1981, 30, for a mention in his diary for February 1848 of dining with 'Hawart'. On Howard's ideas, see Gage 2006, 75.
182. Howard 1838, 75. See also ibid., 77 (on 'shadows being blue or purple'); ibid., 40 (on the 'blue' of 'objects in shadow in the middle and far distance'); and ibid., 101 (on the effect of the 'opposition of positive purple shadows' and of 'contrast' on the light at 'sunset').
183. Ibid., 97.
184. Ibid., 100. See also ibid., 87–88, on the Bianchi's coloured shadows; and ibid., 83, for the argument that the Bianchi 'endeavour to compensate for the want of positive brilliancy, by refining or increasing the delicacy and beauty of the tints', since, 'As we cannot rival the cause' we 'must increase the effect by introducing colour in lieu of those tints which in nature appear neutral'. Howard had been a pupil of Thomas Lawrence.
185. Ibid.
186. See Gascoigne 1997, 33–34, which questions whether the illustrations are chromolithographs at all and not simply hand coloured.
187. Chase 1861, 34.
188. See Cortissoz 1911, 98, for an account of La Farge's visit to the Art Treasures exhibition in Manchester in 1858 and his statement that the Pre-Raphaelites there 'made a very great and important impression upon me' and 'later influenced me in my first work when I began to paint'. Cited in Foster 1979, 8. On La Farge's interest in the Pre-Raphaelites, see also ibid., 5, 8–9, 11, 16–17, 19, and 21 and Yarnall 2012, 42, 80, and 9 n. 127.
189. See Yarnall 2012, 44, which states that the book was lot 42 in the 1866 sale of La Farge's library by Leavitt, Strebeigh & Co. of New York in December 1866. See also Foster 1979, 9 and 11.
190. Cortissoz 1911, 87–88. Cited in Yarnall 2012, 29.
191. See Waern 1896, 12. Cited in Foster 1979, 9.
192. See La Farge 1908, 240, on 'what science could help us in with regard to that light and color' during the early 1860s. Cited in Foster 1979, 9. See also Cortissoz 1911, 112, for La Farge's interest in the 'principles of light and colour' which he applied in 'studies from nature'. Cited in Yarnall 2012, 54. On La Farge's 'optical studies' and his fascination with 'modern scientific analyses', see ibid., 62.
193. See Waern 1896, 27–28; Foster 1979, 21; and Yarnall 2012, 75.
194. Waern 1896, 27.
195. La Farge 1896, 79; citing Fromentin 1859, 257.
196. Cooper 1986, 53 and 79 n. 13.
197. Kobbé 1910, 11. Cited in Cooper 1986, 72.
198. See Tatham 1977, 93 and 97 n. 8; Cooper 1986, 48 n. 16; Hoermann 1988, 105; and Walsh 2008, 199. See also Goodrich 1944, 180 for John Beatty's statement that 'His copy of Chevreul's *Laws of the Contrast of Colour and their Association to the Arts* [was] given to him by Charles in 1873 or earlier'. According to Goodrich 1944, 206, Beatty's recollections of Homer were written in 1923–24 from notes taken in 1903.
199. See Goodrich 1944, 180 and 223. Cited in Tatham, 93; Cooper 1986, 29; Hoermann 1988, 105; and Walsh 2008, 199.
200. Goodrich 1944, 180. For first-hand accounts of Homer's additions, see Tatham 1977, 93–94; Hoermann 1988, 105; and Berrie et al. 2014, 405. The book was Lot 2 in sale 2452 of 15 June 2017 at Swann Galleries, having previously been in the Margaret Woodbury

Strong Museum in Rochester, and before that in the library in Homer's Prout's Neck studio. The same lot contained Homer's second copy, which is unmarked.
201. Goodrich 1944, 180. For Homer's interest in Chevreul's theories, see also Cooper 1986, 74; Hoermann 1988, 105–109; Walsh 1993, 63; and Walsh 2008, 199–205.
202. Vanderpoel 1903, 48–72 and 73–106.
203. Ibid., 63.
204. Ibid., 63–64.
205. Plate CV.
206. Goethe 1840, 363.
207. Ibid. The conversation is recorded in Goethe 1850, I: 345–346.

5 Constancy

Monge and Some Phenomena of Vision

A theory of constancy capable of explaining coloured shadows was, as mentioned at the outset of this book, first set out in Monge's 'Mémoire sur quelques phénomènes de la vision' of 1789.[1] This paper did not actually describe the processes responsible for generating coloured shadows, however, so much as to explain how constancy worked more generally. In fact, Monge only touched on coloured shadows towards the start of his paper, within a *reductio ad absurdam* of the physicalist theory that the 'colours of objects' correspond directly to the 'absolute nature of the rays that objects reflect'.[2] Here, he pointed out that although this theory could explain why the shadow cast by a candle on a sheet of (white) paper 'illuminated' by the 'beautiful blue' of the sky's 'rays' was itself a 'very beautiful blue', it could not explain why the same sheet of paper 'appears white' when the candle is extinguished, even though it is illuminated by the same 'blue' rays.[3] Monge thus undermined the physicalist account of blue shadows he had just aired and left it to his reader to explain them by applying the more general theory he then developed.

An important foundation of this theory was another fact which Monge discussed at the start of his paper, that 'red objects' do not look red when viewed through a 'red glass' as physics would suggest but instead appear 'white' just as white objects do in these circumstances.[4] From this, Monge surmised that 'something mental' must enter 'into the judgements we make about the colours of objects'.[5]

Monge's explanation of the judgement we make in this case is complex and counter-intuitive. It rests on the idea that we estimate the colours of objects on the basis of the fact that all objects – whatever their colour – reflect 'white light' in different quantities.[6] Thus, at one extreme they reflect white light copiously in the form of highlights, while at the other extreme they reflect it in smaller quantities, which are all but indistinguishable from their own local colour. Monge argued that these patterns of reflectance allow us to 'judge the hollows and high points, and in general the degree of inclination of the different parts of the surfaces of bodies'.[7] The same is true, he maintained, of the different amounts of 'white light' reflected by the microstructure of many objects and particularly fabrics, which is made up of 'cylinders' not unlike 'rods of sealing wax'.[8] We therefore see 'white light' reflected from all 'coloured objects', although we are not always aware of doing so.[9]

Monge then pointed out that when we view a scene through a 'red glass', only the 'red rays' in the different amounts of 'white light reflected by coloured objects' can 'pass through' and reach the 'eye'.[10] So it must be these alone, he continued, that 'determine' the 'judgement we make about the inclination of surfaces'.[11] In this situation, therefore, red rays must 'exercise the same crucial function we are accustomed to see[ing] white rays perform', with the consequence 'we are forced take these rays to be beams of white light' and we see red objects as white.[12] We also 'conclude' that red objects look 'white' under these circumstances because they reflect red rays as abundantly as and in the same proportions as 'white objects' do when viewed through a red filter.[13] Although this judgement leads to a spectacular illusion, it indicates more importantly that we habitually use the information about shape provided by the light objects reflect to set the colour of the illumination to white and hence to adjust hues to their true colour. So, although Monge's experiment was conducted under extreme conditions, it shows how we can achieve constancy in normal circumstances, when there is only a weak colour to the light.

The same process of adjustment is demonstrated by another experiment Monge described, which was devised by his colleague Jean-Baptiste Meusnier, and involved illuminating a piece of white paper with 'sunlight' filtered through a 'curtain of red taffeta'.[14] When, subsequently, a beam of unfiltered 'white light' was allowed to pass through a hole in the taffeta, Monge observed, this produced a 'very beautiful green' spot rather than the 'white' spot physics predicts.[15] Similarly, a beam of white light produced a 'very beautiful red' spot when the curtain was green.[16] This happened, Monge explained, because we adjust the 'homogenous' coloured light transmitted by the curtain to 'white' in these circumstances and correspondingly adjust the white light transmitted by the hole to 'another colour'.[17] That is, in subtracting either red or green from the light filtered by the fabric, we subtract the same colour from the light reflected by the spot.

In his conclusion, Monge returned to his opening claim, arguing that the 'judgements we make about the colours of objects do not appear to depend solely on the absolute nature of the light rays that paint their image on the retina'.[18] But this time he explained this fact with a theory of comparison (reminiscent of Scherffer's). He argued, more specifically, that colours 'can be altered by the situation' because 'we are influenced more by the relationship between some of the properties of light rays, rather than the properties themselves, considered in an absolute fashion'.[19] Among these relational properties, Monge listed the comparative 'speed' of different light rays, anticipating our modern idea of frequency.[20] In effect, then, just as Mollon suggests, Monge came close to elaborating the principle that we see constant colours because we compare the amounts of light of each wavelength reflected by the different objects in a scene.

Monge did not apply either theory of constancy to coloured shadows. But, as Mollon has demonstrated, the theory of microcylinders straightforwardly explains how we take a sheet of paper illuminated by bluish daylight to be white. It also implies that we see the shadow on the same piece of paper as blue when a candle is burning with the idea that we adjust its colour in the same direction as we do when we set the colour of the paper around it to white.[21] And for its part, Monge's theory of comparison also explains the shadow's colour as one that we compute by estimating the composition of the light it reflects in relation to that of the light reflected by the other surfaces in the scene.

Monge's observations about coloured shadows, his theory of cylinders, and his experiments with the red glass and the red curtain were all reiterated in the second, posthumous edition of his *Géométrie descriptive* in 1820.[22] Once again, however, while he spelled out how his theory could explain the effects visible with the glass and the curtain, he left it to his reader to work out how it made sense of shadow colour.

Monge's Afterlife in France

The arguments Monge advanced in his 1789 paper were rehearsed by his former student at the École Polytechnique, the engineer, Louis-Léger Vallée, in the *Traité du science du dessin* he published in 1821.[23] Here Vallée began, straightforwardly enough, by setting out Monge's theory that we take the 'yellow' light of a 'candle', the light transmitted by 'panes of coloured glass', and the 'blue light of the air' to be white because these coloured illuminations are reflected by the 'sticks' (or cylinders) in objects in the same informative way as white light is under normal circumstances.[24] Vallée then proceeded to explain the principles at work in the experiments Monge mentioned but in the process gave most of them a slightly misleading twist.

Just as Monge did, Vallée described how we see a spot illuminated by direct 'sunlight' which passes through an opening in 'green curtains' as red.[25] But unlike his teacher, he argued that this happens because white light is 'formed of red and green', so that when we take 'green' light transmitted by the curtain as 'white', the light passing through the opening 'will assume the red colour which is complementary of green'.[26] The problem with this explanation is, of course, that it misrepresents the processes of constancy in terms of the mixture of coloured light which, being an additive process, cannot in any case be explained by the subtractive model he employed.

There is a slightly different problem with Vallée's attempt to make sense of Monge's experiment with candlelight and daylight. In this case, he argued that we take the 'slightly orangey white' of lamp at twilight for 'white', with the result that the shadow it casts will shift towards its 'complementary' colour, 'light blue'.[27] And since the 'shadows' cast by the lamp are already illuminated by the 'azure blue of the light of the air', he continued, they will 'appear a stronger blue' or 'dark blue'.[28] Unfortunately, this account makes several mistakes. Once again, it mischaracterised the colour produced by constancy as a subtractive complementary. Nor is there any reason to suppose that the blueness of the shadow in this situation must be the upshot of a physical process, since it can have this colour when the skylight is grey. Moreover, the light we take to be white in these circumstances is not that of the artificial source on its own, but of the combined illumination formed by both light sources.

Although Vallée rather garbled Monge's theory, he was at pains to distinguish it from the explanations of 'famous Physicists' who did 'not acknowledge' its arguments.[29] Notably he refuted the claim that 'all the phenomena' Monge had discussed could be explained by the theory of 'contrast' on the grounds that they actually had a different 'cause'.[30] He rightly argued, for example, that the effects Monge described appeared 'at first sight' or instantaneously, unlike the 'accidental colours' or after-images produced by 'fatigue' of the 'retina' – as when a 'red square' lying on a white surface' is first fixated and then suddenly removed to produce a 'green' ghost.[31] Vallée was also correct to point out that the colours Monge described were 'of considerable intensity', rather than 'undecided' like the 'opposite' colours produced by viewing a white surface against a coloured ground.[32] He thereby rightly distinguished the process responsible for shadow colour both from successive, and simultaneous, contrast.

Vallée listed the 'colouration of shadows' among the 'causes of visibility' that the artist should imitate.[33] And, for better or worse, his and Monge's ideas on the subject resurfaced in modified form in the *Nouveau traité élémentaire de perspective à l'usage des artistes* of 1823, by Jean-Baptiste-Antoine Cloquet. Although this did not mention Vallée by name, it did mention the 'School' in geometry that Monge directed first at Mézières and subsequently at the École Polytechnique; and it included a section on the 'descriptive geometry' that Monge invented and Vallée popularised.[34]

Cloquet, more specifically, borrowed elements of experiments described by his predecessors for one of his own, which – as shown by a colour illustration (Figure 5.1) – employed a red glass to direct a 'beam of red light' on to a surface in a darkened room 'divided' into bands of white, red, and blue.[35] Unlike Monge, however, Cloquet found that the 'white part of this surface' was coloured red in accordance with the 'tinting power of the ray', just as physics would predict.[36] Cloquet even found to his own surprise that the 'red part' of this surface was not only 'more coloured' than the white part, but also 'more so than the ray itself'.[37] Cloquet conjectured that this happened because the 'red light of the ray' must have 'experienced a second modification' when it encountered the red 'surface', which was akin to the 'initial modification it experienced when it traversed the red glass'.[38] Since, however, red surfaces appear as desaturated as their white neighbours in such situations, his argument is no more than arrant nonsense.

Given that Cloquet did not grasp what was at stake in this experiment, it is unsurprising that there is no trace of Monge's theory in his own, resolutely physicalist, account of coloured shadows. This described how, 'immediately after sunset', the section of an 'opaque white body' illuminated by the warm light of a 'candle' or 'lamp' will 'appear yellow-orange', while its 'shaded' section, 'illuminated by the rays issuing from the azure vault of the sky, will be a 'very deep azure blue'.[39] Similarly, there is no trace of Monge's argumentation in Cloquet's description of how, 'when the earth is covered with snow, and the sky is clear', those 'parts of bodies which are lit by the

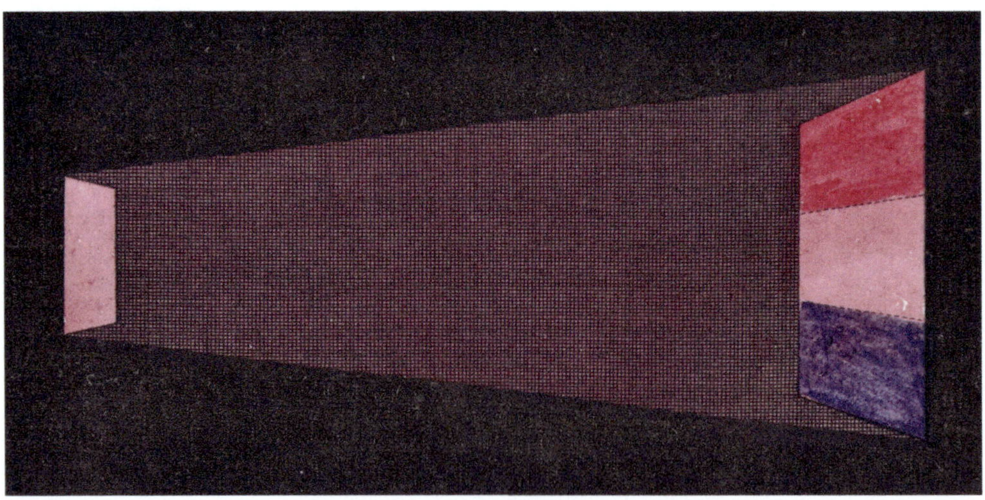

Figure 5.1 Colours under red light. Fig. 7, Plate 8 from Jean-Baptiste-Antoine Cloquet, *Nouveau traité élémentaire de perspective: livre troisième* (Paris: Bachelier, 1823).

Source: Author.

setting sun appear orange, and the shaded parts, as well as the shadows, appear a beautiful azure blue.'[40]

Cloquet's misrepresentation of Monge's ideas did not stop with his book, however, for the unusual reason that the German painter, Johann Erdmann Hummel, lifted the entire section on light and colour in the *Traité* into his book, *Die freie Perspektive* of 1824–25.[41] Hummel did add a footnote to the passage on coloured shadows, but this simply aired Goethe's argument that blue is the colour produced when darkness is viewed through light.[42] Similarly, although Hummel added a section of his own on 'Colour in Shadows', this implied that shadow colour is merely a variant of its local colour, since it recommended making it by mixing the colour of the object concerned with a black made from equal parts of red, yellow, and blue.[43]

As already mentioned (in Chapter 4), Monge's discussion of the experiment involving the coloured light transmitted by a curtain featured in the eclectic exposition of coloured shadows Blanc gave in the *Grammaire*.[44] But although Monge provided a perfectly satisfactory explanation for this experiment, Blanc stated that he did not and offered one of his own based on the Goethean notion that because the eye is made for seeing 'white light', we must complete the light it receives whenever this contains only a 'part' of the spectrum.[45] We consequently see a green spot amid the light transmitted by a red curtain in order to 're-establish equilibrium' in the 'eye fatigued by red'.[46] Implicitly, therefore, coloured shadows are produced by the eye in order to balance out the hue of the coloured light source illuminating the scene. Blanc thus reduced Monge's theory, which could explain coloured shadows as a by-product of constancy, to a metaphysically inspired theory of contrast, which could not.

Monge Abroad

Monge's ideas were in fact taken up earlier in Italy than in France, although with only scant acknowledgement, by the scientist Pietro Petrini. They appeared, more specifically, in a series of publications beginning with a letter from October 1804 published the following year, followed by an article, 'Memoria sopra de' colori immaginarj nell' ombre' in 1805, itself followed by a longer article, 'Ricerche sulla produzione de colori immaginari nell'ombre' in 1807, and ending with the book, *De' Colori Accidentale della Luce* of 1815.[47]

Taking observations made by other predecessors as his starting point, Petrini noted that the shadows cast by the setting sun ran through a sequence from 'violet' to 'green', so that they remained 'complementary' or 'harmonic' in colour to its light as this turned from 'yellow' to 'red'.[48] (And they followed the reverse sequence when the sun rose.) Petrini observed, in addition, that complementary shadows were cast by a light source 'reflected' from a coloured surface or passed through a coloured 'glass' or 'curtain', whenever it acted together with a source which was white.[49] And like Rumford, he accepted that this colour was an 'optical illusion'.[50] When it came to accounting for these facts in his letter in 1804, Petrini resorted to the theory elaborated by Monge – although he attributed this to Meusnier – that the colours we see do not correspond to the 'absolute' quality of the light they reflect.[51] Rather, and shamelessly plagiarising Monge, he argued that our 'judgements' about colour have 'something mental' about them.[52] More specifically, and appropriating Monge once more and Scherffer as well, Petrini claimed that they involve a measure of 'comparison'.[53]

He then made an important suggestion, using a formulation of his own, to the effect that the colour of the shadow in all such cases is generated by our 'subtraction' of any 'excess' colour in the light casting it.[54] And it was perhaps this that caused Helmholtz to cite Petrini's article from 1807 in the bibliography to his *Handbuch der physiologischen Optik* in 1867 that explained coloured shadows in terms of a similar process.[55]

At first, Petrini offered no account of the mechanisms responsible for implementing subtraction. In his article in 1807, however, Petrini did so by enunciating what is effectively a theory of adaptation, which he derived from Scherffer. This maintained that the 'strong impression' made by the light rays from the 'field' around the shadow renders the eye 'insensitive' to the 'weak impression' produced by 'rays of the same kind' reflected by the shadow itself.[56] In his book in 1815, however, Petrini suggested that a better explanation of subtraction could be found in a theory advanced by Pierre-Simon Laplace. This claimed that any 'light' from the 'shadow' of the same colour as the excess in the illumination is 'attracted' by the identically coloured 'rays' reflected by the 'field around' it as they reach the eye – although without explaining how this happens in any detail.[57]

In an attempt to flesh out this contention, Petrini described an intriguing experiment involving a strip of 'canvas' placed over the rear of an opening in a white card.[58] When illuminated by a 'red' light from the front and by the atmosphere from behind, Petrini explained, this appears 'bluish-greenish' in colour.[59] When, however, the surface of the card is coloured red and lit by 'sunlight', the colour of the canvas diminishes. And, as far as Petrini was concerned, this happened because the 'red rays' reflected from the card's surface are now fewer than earlier and so do not take up those reflected by the canvas as much they did before.[60] Needless to say, this explanation is as implausible as Laplace's.

Beginning with the letter he wrote in October 1804, however, Petrini made the valuable observation that the colours he described, including those of shadows, only arose when viewing and comparing two light sources 'simultaneously'.[61] Despite inclining him to conflate coloured shadows and simultaneous contrast effects, this led Petrini to maintain that coloured shadows were not to be confused with the 'accidental' colours Scherffer had described, produced by tiring of the retinal 'fibres' after a 'certain duration of the impression'.[62] Similarly, Petrini made it clear that coloured shadows were not the same kind of phenomena as the 'ocular spectra' or after-images that form after the eye has fixated on a colour 'for some time', which Robert Darwin (Erasmus's son) had described in the *Philosophical Transactions* of 1786.[63]

Monge's experiments with coloured shadows, the red glass, and the red curtain were first described in English in a long review of recent articles in the *Annales de Chimie* published in *The Monthly Review* for 1793.[64] While this summary was in all important respects accurate, the same cannot be said of the analysis Field gave of the experiment with the curtain in his 1811 notebook. Here, under the heading 'Coloured Lights & Shadows', he began by describing a little unclearly how, 'if the Sun shine thru' a *Red or Rose-Coloured* curtain', the '*direct rays*' that 'fall on the opposite wall or a Screen' are 'of a *distinct light Green colour*', while 'at the same time' the '*deflected and dispersed* rays of the sun' that 'pass' through the 'curtain' are 'of a *rose colour*'.[65] Initially, Field speculated that the green colour of the spot illuminated by unfiltered white light was the result of a (mysterious) physical process in which the '*red of the direct rays are* [sic] *reflected and their green transmitted*'.[66] He then conjectured (unlike Petrini) that it was an 'ocular spectrum' of the kind Robert Darwin had described or a complementary after-image produced by 'fatigue' of the 'retina'.[67] And in the end, he speculated that it was 'probable' that the effects produced by the curtain fell under what he called the 'General law of contrasts & of coloured lights & shadows', which stipulated that the 'colours primary & compound of light give shadows of their complementary colours and that the lights passed thro' coloured glasses are invariably as the same'.[68] Field thus reduced Monge's sophisticated theory of constancy to a familiar, catch-all theory of contrast.

The renowned Scottish scientist, David Brewster, referred to the experiment with the curtain in a section in his *Treatise on Optics* of 1831 titled 'On Accidental Colours and Coloured Shadows'.[69] Like Field, Brewster suggested here that the experiment demonstrated the production of 'accidental colours', or 'opposite' or 'complementary colours', in which category he also placed 'ocular spectra'.[70] He also argued that an analogous phenomenon occurred when the shaded 'parts' of a 'white object situated in a 'brightly painted room, illuminated by the sun', assumed a colour 'complementary' to the 'coloured light' reflected by its walls.[71]

At first blush, Brewster was simply wrong to include coloured shadows in the same 'class of phenomena' as 'accidental' colours, just as he was to describe the 'complementary' shadows generated by 'two candles' or 'two holes in a window shutter' when one of them is 'covered with coloured glass' in similar terms.[72] Brewster's position was more complex than this, however, since he insisted that the 'accidental colours' of coloured shadows were 'different' in kind from those of the after-images generated when a portion of the retina is 'deadened' by coloured light, as when a '*bluish green*' appears on a sheet of white paper when a 'red *wafer*' is suddenly removed from it.[73]

He therefore proposed a 'new theory of accidental colours' which was intended to be capacious enough to accommodate its different varieties.[74] The basis of this was that the 'sensation of one colour' is always 'accompanied by a weaker sensation of its accidental of harmonic colour', just as 'every fundamental sound' produces 'its harmonic'.[75] So, for example, whenever we look at 'red', we are 'at the same time, with the same portion of the retina, also seeing *green*', although we do not notice this colour because the sensation of it is 'fainter' than that of the '*red*', which it 'seems only to dilute' by making it 'whiter'.[76] By this account, the green after-image that appears when the red wafer is removed from our sight is nothing other than the 'duration', or conscious persistence, of this previously unnoticed 'harmonic impression'.[77]

The green spot that appeared in the experiment with the red curtain required a more convoluted explanation, however. This began with the assertion that the 'great part of the retina' exposed to the red light transmitted by the curtain is 'in a state of seeing every thing green' unconsciously, while the area of it 'occupied by the image of the white spot' is not.[78] This last portion of the retina nevertheless sees the spot as green, Brewster continued, because the green-producing 'vibrations' at work in the surrounding parts of the retina (somehow) 'spread' over into it.[79] Mercifully, Brewster stopped short of applying this nonsensical explanation to coloured shadows, but had he done so this would suggest their colour were the result of the same mysterious process.

According to F.G. Stephens, Brewster's ideas played a role in Hunt's understanding of 'sunlight effect' and perhaps (since his English is ambiguous) of 'sun shadows' more specifically.[80] It is nonetheless difficult to see what the painter could have made of

Brewster's arguments.[81] Most likely, then, if Hunt really did read Brewster's *Treatise*, he took it to confirm the argument Field and others advanced that coloured shows were produced by contrast.

Goethe and Georg Lichtenberg on White

While Field and Brewster did at least address Monge's ideas about the role of constancy in generating coloured shadows, Goethe chose to ignore them. Indeed, he failed to mention his predecessor's paper in the historical volume of the *Farbenlehre*, although he told Johann Heinrich Voigt in a letter from June or July 1791 that he had 'read' a translation of the 'observations of a French naturalist on blue shadows' – or Monge's 'Mémoire sur quelques phénomènes de la vision' – in Gren's 'Journal der Physik'.[82]

Goethe's obliviousness to Monge's brilliance did not prevent him from approaching a theory of coloured shadows involving a notion of constancy, however. Rather, he did so in the unpublished section of the *Beiträge zur Optik* on coloured shadows, where he proposed that we 'see much more colourful shadows' when they fall 'on a white surface' because a 'completely pure and colourless white surface' is the 'touchstone of all colours'.[83] That is, rather as Monge (and later Land) suggested, Goethe implied that we gauge the colour of shadows in relation to the benchmark objects in a scene we identify as white.[84]

How we see white under such circumstances is nevertheless far from straightforward, as Lichtenberg pointed out in his letter of October 1793. Here, he told Goethe that he could not accept his conclusions because there was 'very much uncertainty in the whole shadow-affair concerning the expressions white, white paper, etc.'.[85] Similarly, Lichtenberg stated in a notebook written between 1793 and 1796 that 'much of what is needed to solve the problem of coloured shadows lies in more precise investigation of what we call "white"'.[86]

At the root of Lichtenberg's objection was his belief that 'we call' surfaces 'white' merely because they have the '*disposition*' to 'reflect all types of coloured lights equally in all directions'.[87] Hence, a surface 'only appears really white' if 'all those coloured rays, in correct proportions, properties and intensities, fall upon it'.[88] This form of words is unfortunate, because it is clear from the context that what Lichtenberg meant is that although white surfaces do indeed look white, they actually present slightly different colours to the eye under different lighting conditions. Thus, they are more or less faintly coloured, objectively speaking, 'in the deepest twilight' and 'in the weakest starlight, by tallow, wax or lamp light, in the brightest sunshine, in the red of evening, by snow and rain, in the woods or in a decorated room'.[89] We nevertheless see these coloured whites as white, Lichtenberg explained, because 'judgement and sensation have so grown together' in our visual experience that 'that we *sense* something which we really only *conclude*'.[90] Lichtenberg believed, in short, that we make an unconscious calculation of some kind about white surfaces, whereby we adjust the actual colour they reflect to white.

Had he pursued this observation, which describes adaptation very well, Lichtenberg might have hit on the role played by white surfaces in generating shadow colour. Instead, however, he tried to do so with reference to the idea that surfaces which looked white were not 'really white' but coloured. He employed this argument in his description of a thought experiment, which involved drilling a hole in the wall of a darkened room to allow the 'varied light' reflected by the coloured houses opposite to produce 'blue, red, yellow, etc., spots' on a white 'sheet of paper'.[91] Here, he claimed, more specifically, that 'coloured shadows arise on so-called white paper' in the same way, 'when some obstruction blocks some of the incident light, which is necessary to bring out the so-called white'.[92] Lichtenberg argued, in other words, that coloured shadows arise when surfaces reflect light sources deficient in one or more of the constituents of white light, just as coloured whites do.

Lichtenberg's commitment to the idea that potentially white surfaces rarely reflect a properly white light led him to suggest that white should not be 'portrayed as white in a painting'.[93] Following this injunction, however, would entail painting colours we do not actually see but those we would perceive if – like an old-fashioned film camera – we registered the wavelengths that reach our retina. Similarly, Lichtenberg's suggestion that the portraitist should 'imagine that in the human face blue, green, yellow and brown coloured shadows can be present' rests on the mistaken assumption that when we see coloured shadows, we apprehend the absolute colours indicated by wavelengths.[94] In short, then, Lichtenberg's potentially insightful argument remained confused and misleading.

Adaptation in Helmholtz and His Predecessors

Petrini's writings not only elaborated a theory of coloured shadows based on Monge's ideas about the role of white light and comparison in ensuring constancy, but they also developed an independent theory premised on adaptation. Another text which did so, slightly beforehand, was the *Theory of Colours and Shadows* by Isaac Milner, which Humphry Repton incorporated into his *Observations on Landscape Gardening* in 1803.[95] This did not elude Field, who recorded in his 1811 notebook:

> [A] short tract on Colour by the Rev.d Dr. Milner Printed in Mr Repton's *Observations on Landscape Gardening* P 214. in which, in the coincidence with my opinion delivered in my *Chromatics* . . . he explains this contrasting colour of shadows (Say Green) by the insensibility of the eye to a weak colour after being powerfully elicited by the same colour with a stronger deeper or more intense state.[96]

As Field suggested, Milner's argument did have affinities with his own views on the origins of coloured shadows in successive contrast, but it was more sophisticated than this. It nevertheless grew, predictably enough, out of the author's puzzlement with the familiar fact that the shadow illuminated by a colourless light source such as 'daylight', the light of a 'small lamp', or a 'wax taper' appears coloured when this acts in concert with the coloured light of 'candles' or a 'lamp' transmitted through a coloured 'glass' or 'reflected' from a coloured surface.[97] And Milner helpfully included an illustration (Figure 5.2) of one such situation in which a lighted taper was employed in conjunction with a large candle or Argand lamp fitted with a red filter to generate red and green shadows.

The premise of Milner's theory was more original, however. This was that 'white light' can not only be thought of as made up from the three 'primary colours' (which he believed to be red, yellow, and blue), but also conceived of as a combination of red and green, or yellow and purple, or any two complementary colours 'called contrasts' (rather as Vallée claimed later).[98] It follows, Milner argued, that the colours of shadows

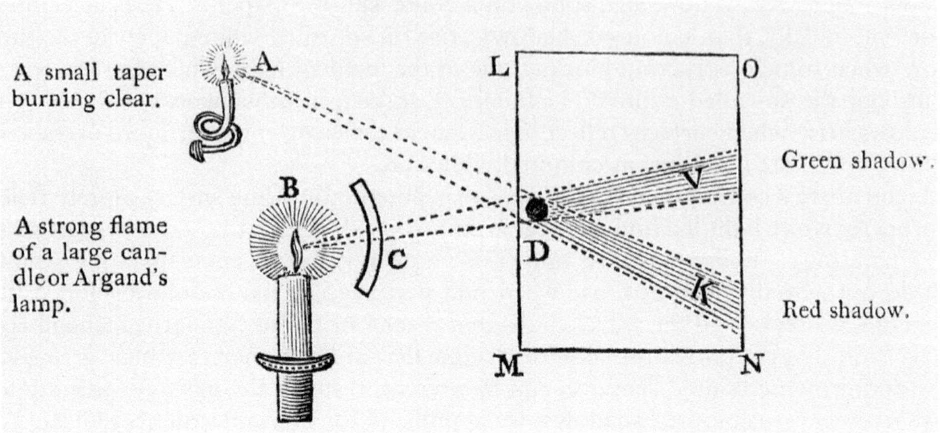

Figure 5.2 Experiment with coloured shadows. Illustration from Isaac Milner, Milner, 'Theory of Colours and Shadows', in Humphry Repton, *Observations on the Theory and Practice of Landscape Gardening* (London: J. Taylor, 1803), p. 215.

Source: Author.

can be understood in terms of the ability of the 'strong' coloured light surrounding a shadow to 'incapacitate the eye from seeing' its opposing colour in the 'dull white light' illuminating that shadow.[99] So, for example, looking at a surround illuminated by a red light source will desensitise the observer to any red in the shadow illuminated by white light that it encloses, and this will appear green. Milner thus articulated a theory of coloured shadows, not unlike Scherffer's, which explained their colour as a consequence of adaptation or one we create by discounting the colour of the illumination.

Having hit on a workable theory, however, Milner took a wrong turn – which may have endeared him to Field – by conflating the process he described with the more familiar process of successive contrast. So, although the phenomenon Milner had described was instantaneous, he likened it to how we 'easily perceive a weak green, or blue' after the eye is 'excited by an intense red'.[100] By this account, the shadow he described appeared green because the 'green' portion of the 'compound' white light illuminating it still 'affects the eye, after the red has ceased to produce any effect'.[101]

Milner supplemented his *Theory of Colours and Shadows* with a 'diagram' or colour circle of sorts (Figure 5.3) which purportedly set out the 'contrasts' that make 'whiteness, or greyness' when mixed.[102] Milner used this diagram to clarify his 'explanation' that 'contrasts' are responsible for the 'appearances' of 'coloured shadows' and to explain that Newton's abortive attempt to make white by mixing coloured powders did 'nothing' to undermine this.[103] In effect, then, he reduced the process responsible for generating coloured shadows to a rather vague theory of colour mixture.

Milner nevertheless believed his theory could explain 'all the difficulties which seem to have embarrassed Count Rumford' and that it could reveal the 'true cause' of the 'blue and green shadows' observed by the 'philosophers' Priestly had mentioned.[104] The verdict of his successors was rather different, however. Repton simply made use of Milner's ideas for his own research into 'harmony'.[105] And John Heaviside Clark only

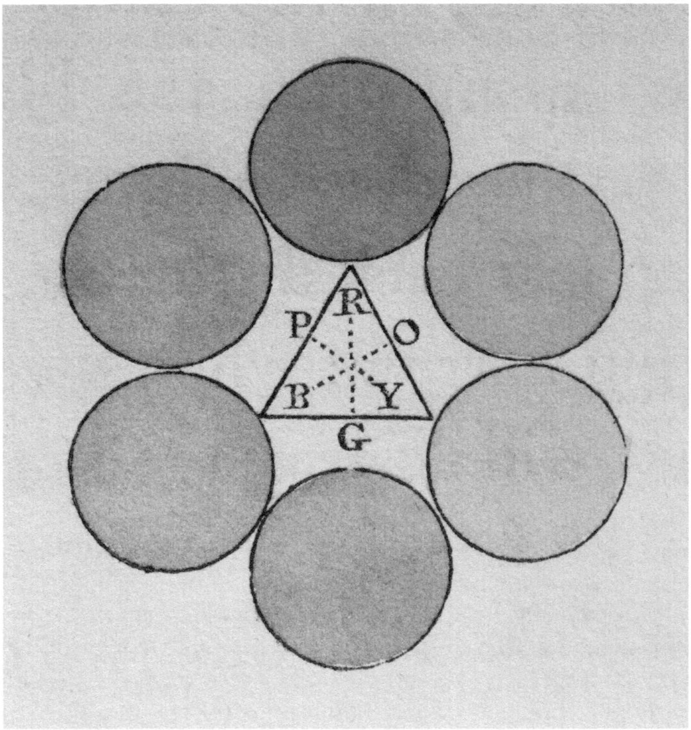

Figure 5.3 Colour circle. Illustration from Isaac Milner, Milner, 'Theory of Colours and Shadows', in Humphry Repton, *Observations on the Theory and Practice of Landscape Gardening* (London: J. Taylor, 1803), p. 219.

Source: Author.

mentioned Milner's text in the section of his *A Practical Essay on the Art of Colouring and Painting Landscapes in Water Colours* of 1807 devoted to 'harmony', ignoring its explanation of coloured shadows altogether.[106]

Although Milner's explanation of coloured shadows had no afterlife, it did find an echo in the theory that Rumford's protégé, Thomas Young, elaborated in the lecture, 'On Vision', that he delivered at the Royal Institution in 1803.[107] Here, like several of his predecessors, Young described how coloured shadows appear when objects are illuminated by a 'coloured light' and another source, including a 'fainter light of the same colour'.[108] He also maintained that the colour of such shadows is always 'opposite' to that of the 'stronger light', arguing that they exhibit the same 'appearances' as the colours produced by what we would call successive and simultaneous contrast.[109]

Notwithstanding, Young set out a theory of adaptation in the continuation of the same paragraph, which explained coloured shadows perfectly. This began with the observation that 'when a considerable part of the field of vision is occupied by coloured light, it appears to the eye either white, or less coloured than it is in reality'.[110] Consequently, Young argued, a 'sheet of writing paper' appears to 'retain its whiteness' even when lit by the 'yellow light of a candle', or the 'red light of a fire' because 'we take away some of the abundant yellow light' from the illumination.[111] In the case of coloured shadows, he continued, the 'effect is nearly the same as if we took away the yellow light from white, and substituted the indico [sic] which would be left'.[112] Accordingly, Young declared, 'in comparison with the light of the candle', the 'common daylight' illuminating the shadow 'appears of a purplish hue'.[113] Young thus developed his argument in the opposite direction from Milner's, from contrast to adaptation. And to that extent, it indicated the cause of coloured shadows more clearly.

One of the few theories of coloured shadows based on a theory of adaptation that did not assimilate them to contrast effects was elaborated by the Inspector of Forests, former *polytechnicien* and painter, Paul Laurent, in his *Traité de perspective aérienne* of 1828.

Laurent began his account in physical terms by describing how 'reflections from the sky' tint the shadows falling on differently coloured natural surfaces, such as wheat and flowers, 'blue', 'violet', and 'green'.[114] He also set out how they lent a 'violet tint' to flesh tones and produced 'greenish' and 'bluish' halftones in the skin.[115] It is nonetheless apparent from his ensuing analysis that Laurent believed they also involved a subjective element produced by our desensitisation to coloured light.

This began by mentioning Buffon's description in a paper of '1754' [sic] of how the colours of the shadows on a white wall towards sunset became 'more and more violet' as the sun tended towards 'yellow and orange'.[116] Laurent departed from his predecessor, however, by suggesting that this happened because the rays of sunlight striking the wall were predominantly 'yellow', and the 'reflections' from the sky illuminating the shadows were mostly 'violet' by which he meant a purple 'rich in red and blue, but poor in yellow'.[117] The shadows assumed their colour, in other words, because the eye became 'jaded' by prolonged exposure to the 'yellow' rays in the 'illuminated part' of the scene and remained sensitive only to those reflected by the shadow.[118] Although Laurent was mistaken in analysing the violet colour of the sky into red and blue components, he was right to suggest that we adjust a slightly yellow light to white, with the consequence that we shift the colour of the shadows it casts in the same cool direction.

Laurent continued by describing how his theory was confirmed by experiments with coloured filters. He maintained, for example, that an object lit by 'yellow' light will cast a 'blue violet' shadow, even 'in the middle of the day', when the light illuminating it is 'blue' pure and simple.[119] And rather as before, he explained that this happens because the eye cannot see any 'yellow' light in the 'shadowed parts' of the scene.[120] Admittedly, his account became quite fanciful when attempting to account for the colours shadows took on; but the principle of adaptation it relied on was perfectly credible.[121]

Elsewhere, Laurent offered some tips to painters on how best to compose the paint colours corresponding to the shadows visible outdoors, including 'Mars violet' for the shadows visible on 'white or coloured objects'.[122] In addition, he illustrated two palettes which contained tones corresponding to 'bluish reflections from the sky' and to 'blue reflections from the atmosphere'.[123] Not content with confining himself to landscape painting, Laurent raised the tantalising possibility of exploiting coloured shadow phenomena for the purposes of interior decoration. Thus, he suggested that if a room were lit by the light passing through a window made of 'red' glass and illuminated simultaneously by white 'blue' daylight passing unobstructed through another window, the resulting shadow would be 'green' – and if the glass were 'green', the shadow would be 'lilac'.[124] Although once again mistaken about the particular colours concerned, this suggestion was sound in principle, and it is perhaps to be regretted that it did not find favour.[125]

Helmholtz, who is widely regarded as the most important vision scientist of the nineteenth century, set out his explanation of coloured shadows in his *magnum opus*, the *Handbuch der physiologischen Optik* in 1867.[126] Like many of his predecessors, he based this both on notions of contrast and adaptation.

The section of Helmholtz's theory which relies on contrast is broadly reminiscent of Goethe's. This argued that when an object on a sheet of white paper is lit by candlelight and grey daylight, the 'red-yellow' shadow illuminated by the candle will appear in 'its objective colouration', while the 'objectively white' shadow illuminated by daylight appear will look 'blue' by dint of 'contrast', which is 'complementary' to the 'pale red-yellow' colour that combined light sources give to the paper around them.[127] And in support of this claim, he pointed out that this blue colour disappears when the candle is screened or when the shadow is viewed in isolation through a reduction tube (as Rumford had discovered).[128] Helmholtz improved on his predecessor, however, by insisting (as Vallée had done) that it is the combined illumination, not the light of the candle on its own, that induces the blue shadow's colour and (as Bossi and Petrini first observed) that this colour arose without the 'assistance of after-images'.[129] Helmholtz also added the entirely original observation that the shadow's colour is 'more vivid' when the eye moves back and forth over it (which suggests it is generated by a process of comparison of the kind described by Monge).[130]

Helmholtz was nevertheless mistaken to claim that 'coloured shadows' are 'most conducive to vividness of contrast' because they fulfil all the conditions required for the effect to take place.[131] It was probably this commitment that led him to gloss over the anomaly that a 'very weak colouration' in candlelight elicited 'contrast colourations' in the shadow it induced every bit as 'intense' as those produced by the 'most saturated' light, so that even a 'weak red-yellow' light engendered a 'very intense blue'.[132]

Notwithstanding, Helmholtz elaborated a second, more powerful theory of coloured shadows. This held that they are generated by a process akin to adaptation (as described by Petrini among others) or how 'we eliminate the colour of the illumination' in a scene without being 'distinctly conscious' of doing so.[133] We do this, Helmholtz explained, entirely by means of unconscious inference from previous experience or by virtue of how 'seeing objects of the same colour' under 'various illuminations' allows us to judge how a 'body would look in white light'.[134] Since the colour we discount in the case of the 'coloured shadows' produced by candlelight and daylight is the 'whitish red-yellow' they produce where the both 'coloured illuminations coalesce', subtracting it renders the objectively white shadow 'blue'.[135] Helmholtz did describe this colour as 'complementary', within a chapter on 'contrast', and to this extent he mischaracterised it. His explanation is nevertheless correct inasmuch as it relies on a theory of adaptation in which constancy is calibrated against familiarity.[136]

Helmholtz disseminated some of his ideas on colour more generally in his essay, 'Optisches über Malerei', originally published in 1876. But although this made no reference to coloured shadows, they are mentioned in the essay, 'Principes scientifiques des beaux arts' by Brücke, which appeared alongside the French translation of Helmholtz's essay in 1878.[137]

Brücke had in fact stated in his *Die Physiologie Der Farben* of 1866 that 'contrasts which are very striking and interesting for the artist and scientist can be observed in experiments involving twin light sources' when one of these is coloured.[138] And he described, more particularly, how the shadow of an object lit by neutral 'daylight' in a scenario of this kind 'will appear tinged with blue' or the 'colour contrasting with the background' illuminated by both sources.[139] Brücke also mentioned here how the 'coloured light of the sun, at sunrise and sunset' sometimes 'gives rise to the most grandiose contrasts', including the 'marked blue shadows' that 'sunset projects on a snowy surface'.[140]

In his later essay, Brücke offered a detailed discussion of the coloured shadows produced by a combination of 'lamplight' and 'daylight'.[141] In this, he argued that the shadow cast by the artificial source was 'blue' because the daylight illuminating it contained more blue than the 'dominant light', itself a 'mixture' of both sources.[142] It thus appeared 'distinctly blue', Brücke continued, 'by contrast' with the warm light illuminating the 'supposedly white' paper.[143] While this move was retrograde, his next move was highly original, which was to argue that the shadow cast illuminated by the candle appeared 'brownish' by the same principle.[144] Brücke claimed, that is, that it appeared 'distinctly brown' rather than 'orange', because it both contained more 'red and yellow' than the combined illumination and was darker than the white background on which it fell.[145] Although Brücke assimilated this effect to contrast, he did in fact identify the muddying effect that constancy has on this kind of shadow.[146]

Colour for Philosophers

No account of the assimilation of coloured shadows to contrast is complete without some reference to C.L. Hardin's *Color for Philosophers* of 1988.[147] Here, for example, in his analysis of experiments involving 'two sources of light with somewhat different spectra', such as 'candlelight and daylight, or candlelight and incandescent light', Hardin argued that the shadows produced are 'contrasting colours, usually magenta (a bluish red) and a bluish green'.[148] He explained these pairings with the argument that the 'extra bluish component in both shadows will depend on the extent to which the light sources are yellowish rather than chromatically neutral'.[149] This claim is difficult to credit, however, since shadows in scenes of this kind simply do not have the colours Hardin describes. Moreover, it remains obscure how yellow in the warm light source can induce a bluish tinge in the shadow it illuminates.

Hardin then described the coloured shadows produced by overlapping the beams of two projectors, one emitting uninterrupted white light and the other the coloured light transmitted by a 'piece of transparent green celluloid' overlaid with a 'piece of opaque tape' in the form of a cross.[150] Hardin pointed out that although this cross casts a shadow whose absolute or physical colour was a 'slightly yellowish grey' (presumably from the unfiltered light from the first projector), we nevertheless see a 'bright *pink* cross on a greenish field'.[151] This result, he argued, is 'entirely the product of the chromatic opponent system' or of the contrast between the colour of shadow and the slightly desaturated, yellowish 'green' of its surround.[152] Hardin also maintained that this was proven by the fact that the shadow's colour reverted to that of the flux illuminating it when viewed through a reduction tube.[153]

Having established this much to his own satisfaction, Hardin added that 'With ingenuity and photographic skill', the 'effect', by which he meant the production of coloured shadows by simultaneous contrast, 'can be tricked up to be far more impressive, and this seemed to be what happened with Edwin Land's celebrated two-projector demonstrations'.[154] The effects at issue were in fact produced by computations of the kind involved by constancy.[155] And, as Hardin conceded, they led Land to develop his Retinex theory.[156] Notwithstanding, Hardin insisted that Land 'soon acknowledged' that these 'two-filter effects were, at root, coloured shadow-phenomena, described in some detail by J. W. Goethe' and 'even earlier, by Count Rumford'.[157] Hardin thus effectively accuses Land of mistaking the operation of simultaneous contrast in producing coloured shadows for a constancy effect, when in fact the opposite is true, and it is Hardin who misunderstands their causes.

Hardin did nonetheless admit that it can be 'tricky' to catch the effect of 'simultaneous contrast' when it comes to accounting for 'blue shadows' of the kind the Impressionists painted, which appear on snow when the 'yellow rays of the sun are interrupted'.[158] He explained, however, that this is because it is normally a different – and purely physical – process which produces shadows of this kind, namely that 'Snow reflects' the 'blue light' scattered by the 'atmosphere', which remains visible 'in shadows' where the 'yellow sunlight is blocked' and cannot cancel out its 'perceptual effect'.[159]

Although Hardin maintained that simultaneous contrast does not produce coloured shadows in these circumstances, he argued that it does in another situation. This happens, he claimed, when a 'large cloud' passing 'directly overhead' blocks out the blue 'skylight' without 'completely obstructing the sun', thereby allowing just enough yellowish sunlight through to make 'cast shadows' appear 'true blue' by 'contrast' with the snow that it tinges this colour.[160] One problem with this argument is that contrast, as already mentioned, is too weak to produce the effect described. Sunlight is also white or only faintly yellow in the daytime. So, for all its apparent sophistication, Hardin's explanation makes no significant advance on Lostalot's or Vandepoel's. It is regrettable, therefore, that commentators continue to maintain that contrast is responsible for creating the coloured shadows visible in nature and depicted in Impressionist painting.[161]

Notes

1. For an analysis of this text, see Mollon 2003, 20–21 and Mollon 2006. For a translation with a commentary, see Kuehni 1997.
2. Monge 1789, 135.
3. Ibid., 133–136.
4. Ibid., 132–133. Leonardo expressed the common-sense view in Chapter 162 of the *Trattato*, where he claimed that: '[R]ed glass makes whatever appears behind it appear red'. See Farago et al. 2018, 2: 720; da Vinci 1540, f77r; and da Vinci 1651, 77r. See Benson 1895 (an unpublished letter of October 1895 placed inside a copy of Benson 1871) for how the author explained 'WHY SCARLET & RED LOOKED AT THRO' RUBY GLASS APPEARS GREY OR CREAMY' with the argument that 'even the brightest ruby glass, besides quenching nearly all the other prismatic rays, actually weakness the red rays themselves down to perhaps less than half their proper intensity'.
5. Monge 1789, 135. See also ibid., 133–134, which notes that when a glass allowed only 'homogeneous rays of a certain type' (or particular colour) to pass through it, 'bodies which reflect only this type of rays' should 'appear white'. Monge noted nevertheless that although this did happen with a 'yellow' glass, it was not so clearly the case with 'blue, green, and violet' glasses.
6. Ibid., 137.
7. Ibid., 138.
8. Ibid. See Mollon 2006, 302, which argues that Monge came close to anticipating the 'chromatic convergence hypothesis' in this discussion. This stipulates that the ratios between the amount of its own ('body') colour and ('specular') highlights reflected by different parts of an object can be plotted on a line which theoretically terminates at the point where only the specular component of the illumination is present and thereby specifies the colour of the illuminant. On the chromatic convergence hypothesis, see Hurlbert 1998, 310–312. For the argument that 'specular highlights' contribute to constancy by other routes, see ibid., 312–313 and Foster 2011, 676–677.
9. Monge 1789, 142.
10. Ibid.
11. Ibid.
12. Ibid., 142–143.

13. Ibid., 143. See also Mollon 2006, 302, for the explanation that red objects in the scene Monge described look white because they imitate a property which is normally unique to white objects, namely that there is no variation in their hue under a uniform, and putatively white, illumination because their body colour and their 'specular' highlights are the same (white) colour. Scherffer had made a similar point, as mentioned earlier.
14. Monge 1789, 145–146.
15. Ibid., 146.
16. Ibid.
17. Ibid.
18. Ibid., 146.
19. Ibid., 146–147.
20. Ibid., 145.
21. Ibid., 135–136.
22. Monge 1820, 184–187. The relevant section of this text, titled 'Théorie des ombres et de la perspective', was 'extracted' from the '*Leçons inédites de M. Monge*' by his former pupil and nephew by marriage, the engineer and mathematician, Barnabé Brisson. See also Brisson 1818 for Brisson's obituary of Monge.
23. Vallée 1821, 343–353. See also ibid, 341, for a short, physicalist account of the 'blue' shadows illuminated by the sky, as described by Leonardo, and of the 'green' shadows the sky can sometimes produce.
24. Ibid., 349.
25. Ibid., 351.
26. Ibid.
27. Ibid.
28. Ibid. See also ibid., 351–352, for the implausible argument (not unlike one of Field's) that 'clouds' at 'sunset' appear 'orangey', although they reflect only 'white light', because we take the 'pale blue' of the sky to be 'white' and adjust their colour accordingly.
29. Vallée 1821, 351, referring to the refutation of Monge's theory of constancy and his ideas on coloured shadows in Prieur 1805, 15–25.
30. Ibid., 412–413.
31. Ibid., 412.
32. Ibid., 413.
33. Ibid., 372. This remained the case although Vallée's book was republished in 1838.
34. Cloquet 1823, 213 and 71–154. For their writings on descriptive geometry, see Monge 1820 and Vallée 1819. The title page of Vallée 1821 describes itself as following on from this earlier book. As Professor of Drawing and Perspective at the École des Mines during the revolution, it is likely Cloquet knew Monge, who was a prominent Jacobin. On Monge's politics, see Mollon 2006, 307–308.
35. Cloquet 1823, 200.
36. Ibid.
37. Ibid. Here, Cloquet also suggested that the red beam coloured 'the blue part' of the surface 'purple' because it 'must have been modified towards blue when it met this surface'.
38. Ibid.
39. Ibid., 169–170.
40. Ibid., 170.
41. See ibid., 154–203 and Hummel 1825, 2: 92–95 (esp. 27–29 and 85–86).
42. Hummel 1825, 2: 27–28.
43. Ibid., 92–95.
44. Blanc 1867; citing Monge 1820, 184–185.
45. Blanc 1867, 600.
46. Ibid.
47. Petrini 1805 (January–February); Petrini 1805 (May–June); Petrini 1807; and Petrini 1815.
48. Petrini 1805 (January–February), 206–211; Petrini 1805 (May–June), 382–384; Petrini 1807, 39–40 and 42; and Petrini 1815, 4–5, 12–13, and 16. Petrini stated that the word 'complementary' had been coined by Hassenfratz (in Hassenfratz 1802) rather than Rumford. On the history of the word, see Roque 1994, 412–414. Petrini explained further that a colour is complementary to another when they lie in corresponding portions of the two halves of the spectrum.

49. On reflected light, see Petrini 1805 (January–February), 212–213; Petrini 1805 (May–June), 377–379; Petrini 1807, 43–46 and 53–55; and Petrini 1815, 19–22. On light filtered through coloured glass, see Petrini 1805 (May–June), 388–390; Petrini 1807, 54–55; and Petrini 1815, 25–30. On light passed through a curtain, see Petrini 1805 (January–February), 213–214; 1805 (May–June), 381–382; 1807, 48–49; and Petrini 1815, 20 and 50.
50. 1805 (January–February), 214; Petrini 1805 (May–June), 391–392; Petrini 1807, 47; and Petrini 1815, 22.
51. Petrini 1805 (May–June), 380–382 (citing Mesunier); and Petrini 1807, 49–50. See Petrini 1807, 49 and Petrini 1815, 50–51, which cite Monge's account of Meusnier's experiment with the curtain. See also Petrini 1815, 49 for another passing mention of Monge.
52. Petrini 1805 (May–June), 382; and Petrini 1807, 49 (which footnotes Monge).
53. Petrini 1805 (January–February), 214; 1805 (May–June) 379–380 and 385; and Petrini 1807, 48 and 51. For more on both ideas, see Petrini 1805 (May–June), 388 and 390, 392–393; Petrini 1807, 39, 44, 46, and 56; and Petrini 1815, 39.
54. Petrini 1805 (January–February) 211 and 214; 1805 (May–June), 380, 392, and 394; and Petrini 1807, 42, 44, 46, and 56–57.
55. Helmholtz 1962, 2: 298. This misdates it to 1783.
56. Petrini 1807, 44. See also ibid., 48 and 62. This argument draws on Scherffer 1765, 12–13.
57. Petrini 1815, 43–44. Citing Laplace as paraphrased in Haüy 1806, 2: 271.
58. Petrini 1815, 44.
59. Ibid.
60. Ibid., 45.
61. Petrini 1805 (January–February), 211 and 215; 1805 (May–June), 388 and 391; Petrini 1807, 42 and 54, 56, 66; and Petrini 1815, 14. Petrini stated that he had discovered this fact before Prieur published his findings on viewing two colours simultaneously. See Petrini 1807, 37 and Petrini 1815, 19 and 62. On Prieur, see also Petrini 1807, 3, 38, 43, 49 and Petrini 1815, 66.
62. Petrini 1807 62–64.
63. Petrini 1815, 58. On 'ocular spectra', see Darwin 1786, 317–318.
64. Anon 1793, 517–518. The first English translation of the *Géométrie descriptive* including the sections on coloured shadows was not published until 1851.
65. Field 1811, f98r.
66. Ibid., emphasis in original.
67. Ibid.
68. Field 1811, f98r.
69. Brewster 1831, 304–312. There is no evidence that Brewster knew Field's notebooks.
70. Ibid., 304–308. See ibid., 305, for the explanation that '*complementary* colours' are those which 'when mixed together, make white light' and which lie 'opposite' one another on a colour 'circle'. This repeats the argument advanced in Newton 1704, 116, that mixing rays of light lying 'opposite' one another on the colour circle creates white light.
71. Brewster 1831, 308.
72. Ibid., 309.
73. Ibid.; referring to ibid., 304–05.
74. Ibid., 309.
75. Ibid.
76. Ibid.
77. Ibid., 310.
78. Ibid.
79. Ibid.
80. See Stephens 1860, 20. On Brewster and Pre-Raphaelitism, see Glanville 2004, 30. See also Gage 1989, 65, for the suggestion that Hunt was led to Brewster by reading David Ramsay Hay and Hunt 1905, 29, for the artist's recollection that his father had given him a copy of 'Harmony of Colour' by 'Hayes', by which he probably meant Hay 1828 (or a later edition of the same text).
81. Doubt is cast on Stephens's account by his suggestion that Hunt also read Newton and Humphry Davy on the appearance of sunlight and coloured shadows, since neither author discussed these. On the Pre-Raphaelites and Newton, see Glanville 2004, 30. For Davy's text on ancient pigments, see Davy 1815.

82. See Goethe 1887–1912, 9: 274.
83. Goethe 1897, 121–122. The translation is Monge 1790.
84. See also Roux 1824–29, 3: 77, which argued that candlelight in a room lit by daylight appears more 'yellow' than normal when 'surrounded' by a 'white surface' and conversely that daylight under the same conditions assumes a 'bluish-grey' colour.
85. Joost et al. 2002, 302.
86. Lichtenberg 2012, 161.
87. Joost et al. 2002, 302.
88. Ibid.
89. Ibid.
90. Joost et al. 2002, 302.
91. Ibid.
92. Ibid.
93. Ibid., 302.
94. Ibid.
95. See Repton 1803, 214, on how he received the text from William Wilberforce, who had described Milner as the 'ingenious and philosophical friend, who himself can paint very well'. See also Repton 1816, 49–51, for a later text, headed 'Concerning Colours', addressed to Wilberforce, in which Repton alluded to Milner's *Theory of Colours and Shadows*. On Milner's essay, see Finley 1967, 362 n. 32.
96. Field 1811, f73v.
97. Milner 1803, 215.
98. Ibid., 215–216. On the complementary wavelengths that make white, see Hurvich 1997, 84–85.
99. Milner, 216.
100. Ibid., 217.
101. Ibid.
102. Ibid., 218. See also Finley 1967, 362.
103. Milner 1803, 218–219.
104. Ibid., 217.
105. See Repton 1803, 220–222, for the author's opinion that Milner's ideas were useful to the theory of '*harmony*' in 'art' and 'nature' and Repton 1816, 49–51, which used Milner's ideas as the basis of a discussion of harmony.
106. Clark 1807, 15. On Clark and Milner, see Finley 1967, 362.
107. On Young and Rumford, see Wood 2011, 118–121.
108. Young 1807, 1: 456.
109. See ibid., 455 for descriptions of the coloured after-images produced by successive contrast, as when 'the eye has been fixed on a small object of a bright colour and is then turned away to a white surface' and of the supposedly 'similar effect' (actually produced by simultaneous contrast) when a 'white or grey object is viewed on a coloured ground, even without altering the position of the eye'.
110. Ibid., 456.
111. Ibid.
112. Ibid.
113. Ibid. Young's use of 'purplish' should be regarded as casual, since he had already developed an additive theory of light mixture, according to which the complementary of yellow was blue. On this, see Young 1802, 395.
114. Laurent 1828, 67–68.
115. Ibid., 71–72.
116. Ibid., 73.
117. Ibid., 73–74.
118. Ibid., 74.
119. Ibid.
120. Ibid.
121. Ibid., 74–75.
122. Ibid., 93. See also ibid., 99–100, for Laurent's advice on which pigments to use for the lights when shadows are 'violet' and the sun is 'low on the horizon'.
123. Ibid., 112–113.
124. Ibid., 75.

125. Perhaps this was partly because the book was self-published, as was the second edition of 1838; although a revised edition of 1840 did have a publisher (Librairie scientifque-industrielle L. Mathias).
126. For a summary of Helmholtz's ideas on coloured shadows, see Schöller 2017, 22–29.
127. Helmholtz 1962, 2: 271.
128. Ibid., 217–272. Cf. Rood 1879, 256, which argues that 'the illusion . . . even after the causes which gave rise to it have disappeared . . . still persists', that is, when the shadow is viewed through the tube.
129. Helmholtz 1962, 2: 271, which also states that the colour is 'more vivid' when the eye moves back and forth over it.
130. Ibid.
131. Ibid., 271.
132. Ibid., 276.
133. Ibid., 287.
134. Ibid.
135. Ibid. See ibid., 274, for the argument that when a 'particular colour' is 'dominant in the visual field, a paler shade of the same hue will look white to us, and real white will seem to be the complementary colour'. See also ibid., 276.
136. On the role of familiarity and memory in compensating for illuminant colour, see Foster 2011, 689–690.
137. Brücke and Helmholtz 1878, 9–168.
138. Brücke 1866, 179.
139. Ibid., 179–180. Brücke also mentioned the 'magical effects' produced by the 'white and coloured light' which passes through stained glass and clear windows when both are present together.
140. Brücke 1866, 180. Cited in Roque 1996, 36 and Roque 2009, 306.
141. Brücke and Helmholtz 1878, 106–107.
142. Ibid., 107.
143. Ibid.
144. Ibid.
145. Ibid.
146. Goethe, as mentioned earlier, referred to 'yellowish-brown' shadows illuminated by the stringer of two light sources. See also Thompson 1794, 115 and 117, for a description of 'brown' shadows and Joost et al. 2002, 336, for Lichtenberg's description of a 'dirty yellow' shadow.
147. Hardin 1988, 50.
148. Ibid.
149. Ibid.
150. Ibid.
151. Ibid.
152. Ibid.
153. Ibid.
154. Ibid.
155. See Land 1959, 289–290, for an account of the experiments concerned, which demonstrated how we see the full range of colour (albeit a little washed out) when two black and white slides of the same view, made through filters transmitting long-wavelength ('red') and medium-wavelength ('green') light, respectively, are projected and superimposed, provided only that the light emitted by the projector containing the 'red' slide is of a longer wavelength than the other.
156. Hardin 1988, 187–193.
157. Ibid., 188.
158. Ibid., 50–51.
159. Ibid., 51.
160. Ibid., 50; paraphrasing Walls 1960, 35 n. 4.
161. See Charnay and Sieffert 2001, for the argument that the shadows in Monet's *The Magpie* result from causes first explained by Rumford and Chevreul and Schöller 2017, 97 which compares the effect Monet observed to the contrast effect Goethe considered responsible for the blue shadows he saw in the Harz mountains.

6 The Criterion

The Verbal Sketch

Although science inclined some artists to overlook coloured shadows where they were visible, and others to see them where they were not, their paintings and verbal reports indicate that many could still perceive them in their full vividness. It remains the case, however, that the first artists to engage with these fugitive phenomena were hampered by the lack of media in which they could record them quickly. Most importantly, stiff, quick-drying oil paint which allowed artists to fix effects they saw on the spot was not available until the 1860s, making it difficult to capture them *in situ* in a painting of any size. There are consequently few works which do so before this date, many of which are small watercolours. Carl Blechen made several of these in Italy in the 1830s; but he also made a number of remarkable, tiny oil sketches, including *Courtyard With Arches and Fountain* of 1829 (Figure 6.1), which gives striking expression to the shadow colours visible at sunset.[1]

In situations where visual phenomena are evanescent and mutable, knowledge and belief become crucial to shaping our visual experience. Carus, for example, remarked in a letter of around 1820: 'Countless beautiful effects of refracted colour present themselves to the *practiced* eye'.[2] And judging from the accounts given by the earliest painters of coloured shadows more specifically, their ability to verbalise their experience played a major role in shaping what they saw. Indeed, it is clear from the descriptive richness of several artists' words, and from the comparative poverty of their paintings, that a statement sometimes served better as what Wittgenstein called the 'criterion' – or public expression – of a 'visual experience'.[3] This is sometimes true of Carus, whose verbal sketches combine a poetic turn of phrase with an acuity of observation consistent with his training as a scientist. In his letter of around 1820, for example, he described how:

> On these glorious winter days I have been spending much of my time in the open air, delighting in the varied and exquisite interplay of light between blue sky and snow-covered earth. . . . There, a bright gleam of snow-light sparkles on a rocky edge . . . here, snow-drifts lie in shade, and the raised parts of their surfaces show as bluish, occasionally purplish tones.[4]

By comparison, Carus's *Cemetery on Mount Oybin* of 1828 (Figure 6.2) is quite prosaic. It may even falsify the phenomenology of coloured shadows inasmuch as flecks of blue paint on the right arm of the cross, on the top right corner of the tallest section

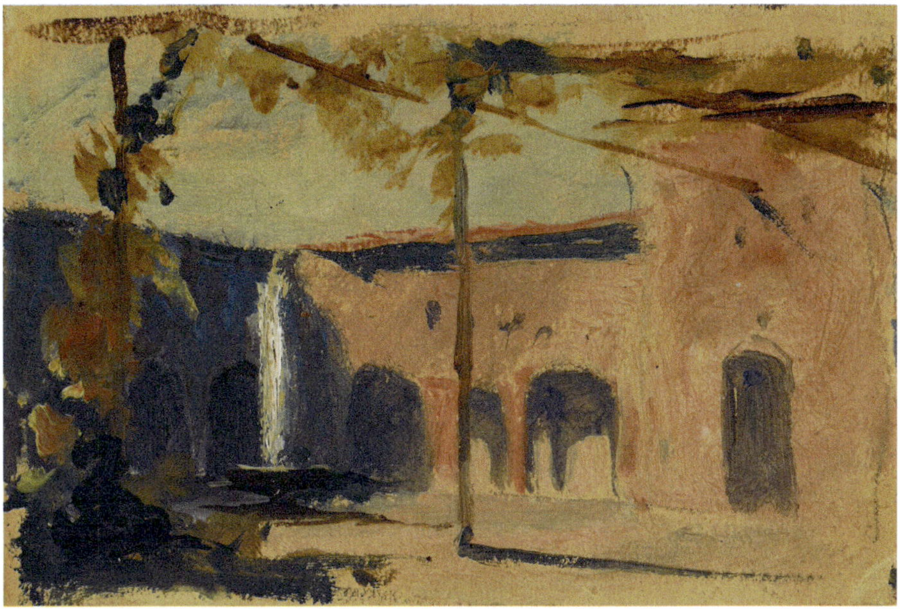

Figure 6.1 Carl Blechen, *Courtyard With Arches and Fountain*, 1829. Oil on paper. 6.8 × 10 cm. Berlin, Kupferstichkabinett, Staatliche Museen zu Berlin, inv. no. SZ Blechen 232.

Source: Dietmar Katz.

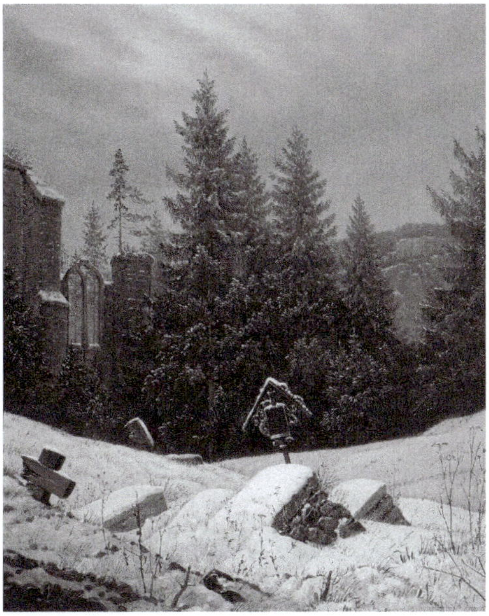

Figure 6.2 Carl Gustav Carus, *Cemetery on Mount Oybin*, 1828. Oil on canvas, 67.5 × 52 cm. Leipzig, Museum der Bildenden Künste.

Source: MdbK/AKG Images.

of wall, and in the snow below this imply that they are visible on a cloudy day.[5] If so, then perhaps Carus may have added effects he observed in bright but unsettled conditions to those he could more easily observe and note down, under a dull but stable illumination.

Samuel Palmer gave an eloquent if eccentric account of coloured shadows on the first page of his 1824 sketchbook. He described the tightly intercalated patches of warm sunlight and cool shadow visible on the Gothic buildings around Westminster Hall from across the Thames on a July evening at nine o'clock in the evening, as follows:

> [T]he melting of many gradations in the lights opposed to elaborate shadows & architectural members near by it the sharpness of the cool shadows against the cool sky &c could I should think give that mild glimmering poetical light of eventide.[6]

Even though the drawing is in black and white, it may be revealing that this description takes over from the sketch Palmer made of the scene on the same page (Figure 6.3) at the very point where this stops. Palmer had to be content, at all events, with describing the problems he would have encountered, if only he could have represented the scene in paint. These are mostly concerned with the control and manipulation of contrast effects in the picture and include the suggestion that the 'less elaborate building with strongly marked shadows would through a neutral tint bear out against a flat mass of the most vivid colour'.[7]

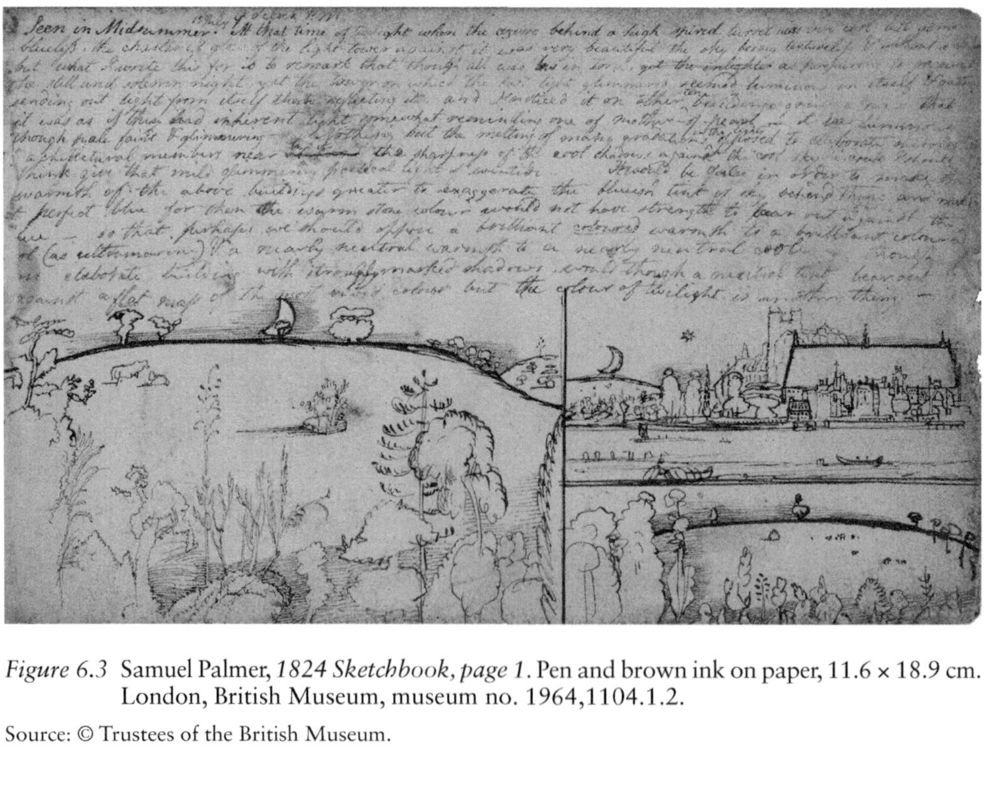

Figure 6.3 Samuel Palmer, *1824 Sketchbook, page 1*. Pen and brown ink on paper, 11.6 × 18.9 cm. London, British Museum, museum no. 1964,1104.1.2.

Source: © Trustees of the British Museum.

Another passage in the 1824 sketchbook evokes a particularly subtle effect in the form of the shadows generated in the earth by areas of backlit grass. This reads:

> Looking at grass with my face to the sun, he shines through each blade making masses of the most splendid green and yet intimately warm so warm that we can only liken it to yellow & yet most vivid green . . . where . . . earth in shadow (quite purple) comes immediately against the grass in transmitted light the effect is indeed splendid.[8]

And on the following page of the same notebook, Palmer recorded how:

> We observed the shadows in Greenwich Park in the morning so purple & cool that had the colour of the lights been also laid over where the shadows were to come, the shadow colour would exactly have been made with bold touches of purple made of brilliant blue & lake.[9]

Again, however, no surviving work gives shape to the coloured shadows Palmer described so well.

Brown also experienced great difficulty in fixing effects of sunlight and shadow, despite employing a quick-drying 'copal' medium to enable him to paint more rapidly.[10] Indeed, Brown explicitly bemoaned the impossibility of capturing coloured shadows in paint in a letter of July 1855, which probably relates to the painting that became *The Hayfield*. This moving document reads:

> I want little subjects that will paint off at once. How despairing it is to view the loveliness of nature towards sunset & know the impossibility of imitating it, at least in a satisfactory manner as one could do would it only remain so long enough. . . . What wonderful effects I have seen this evening in the hay fields, the warmth of the uncut grass, the greeny greyness of the unmade hay in furrows or tufts, with lovely violet shadows & long shades of the trees thrown athwart all & melting away one tint into another imperceptibly, & one moment more & could passes & all the magic is gone. Begin tomorrow morning all is changed . . . & all that was lovlyest [sic] is all that is tamest now, alas! It is better to be a poet – still better a meer [sic] lover of nature, one who never dreams of possession.[11]

Despite the impressive articulacy these artists achieved, the limitations inherent to language meant that their words could not capture the fine-grained particularity of their visual experience. Brown was not wrong, then, to lament the poverty of the resources available to him.

The difficulties presented by the perceptual elusiveness of coloured shadows were not entirely insuperable, however. For one thing, the process of making a painting of coloured shadows, however primitive or unsatisfactory, could alert an artist to what she or he had seen even after the fact, and thereby rescue this from oblivion. In such cases, the work of art comes to serve as a 'criterion' of a visual experience rather as sometimes it is only when we find the appropriate form of words that we realise what we are thinking.[12] Paintings such as *Carrying Corn* (Figure 6.4), which Brown painted over at least 7 months beginning in the summer of 1854, attest to the power of painting a scene over and over to do this.[13] It gives expression, more particularly, to the

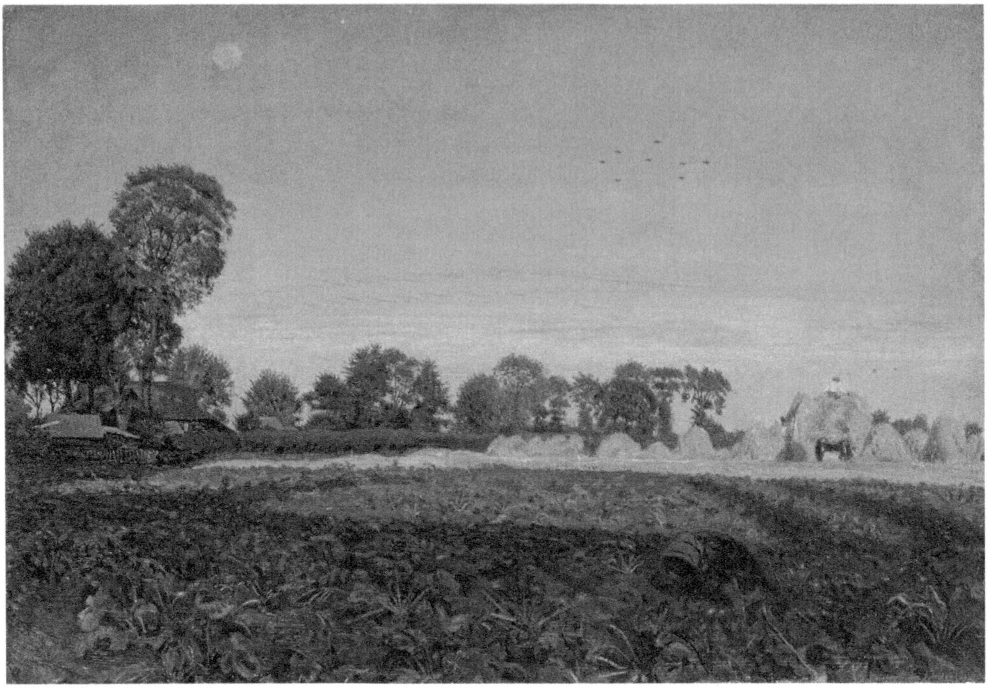

Figure 6.4 Ford Madox Brown, *Carrying Corn*, 1854–55. Oil on mahogany, 19.7 × 27.6 cm. London, Tate Britain. Tate reference no. N04735.

Source: © Tate.

blue and violet shadows that appeared fleetingly towards evening in the turnips in the foreground and in the buildings and trees in the background.

Framing in Carus and Friedrich

A painting can render our visual experience present especially well when it forms part of a system which involves a particular viewing technique capable of facilitating this result, often with the aid of a device. Among the most important of such techniques and devices were those that cropped, or framed, the motif. Most likely, painters originally employed these fortuitously with different intentions. Some undoubtedly did so as a result of following Valenciennes's advice to view the landscape through a 'window whose opening serves as a frame' to establish its 'composition'.[14] Others achieved the same result by working within a shaded section of the landscape, or a chamber of sorts, with the intention of making the picture under subdued lighting similar to the interior light under which it would eventually be displayed (thereby compensating for the fact that it looks duller with less contrasting indoors).[15]

A clear example of this kind of framing occurs in Carus's *A River Cruise on the Elbe River Near Dresden* of 1827, which presents the sunlit scene outside from the position of a spectator located within the shade of the barge's canopy. The painting's small size

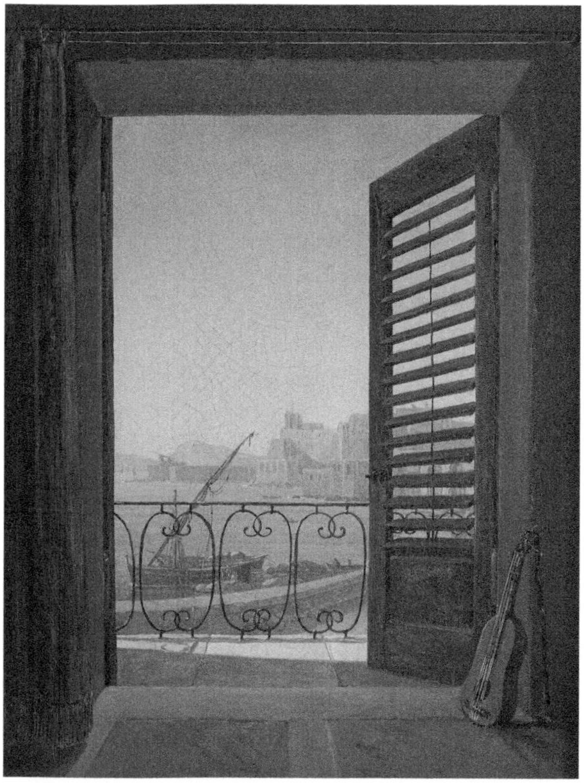

Figure 6.5 Carl Gustav Carus, *Balcony Room With a View of the Bay of Naples*, *c*. 1829. Oil on canvas, 28.4 × 21.3 cm. Berlin, Nationalgalerie, Staatliche Museen zu Berlin, inv. no. FNG 59/92.

Source: Staatliche Museen zu Berlin/AKG Images.

suggests that Carus began it out of doors or really did depict what he observed from the vantage point it implies. In any event, framing seems not only to have enhanced the strong colour of the shadows on the oarsman's shirt and waistcoat, but also to have intensified the perceptually less obvious blue colour in the shadows on his companion's dress. A similar situation occurs in Carus's *Balcony Room With a View of the Bay of Naples* of 1829 (Figure 6.5) in which the shadows on the pavement outside, like those on the walls and the two boats behind it, appear distinctly violet–blue to the spectator who looks out at them from inside the room depicted.

A possible model for *A River Cruise on the Elbe River Near Dresden* is *Chalk Cliffs on Rügen* of 1819 (Figure 6.6) by Friedrich, whom Carus had met in Dresden in 1816 before becoming friends with him around 1818.[16] Although Friedrich did not make easel paintings directly from nature, the composition of the work implies that the artist viewed the white Stubbenkammer cliffs at some point, perhaps while visiting Rügen most recently in January 1818, from within the darker space framed by the trees on either side of the scene.[17] Blue shadows, lit by the sky, are clearly visible on the white chalk, and alongside these are dirty yellow–orange shadows illuminated

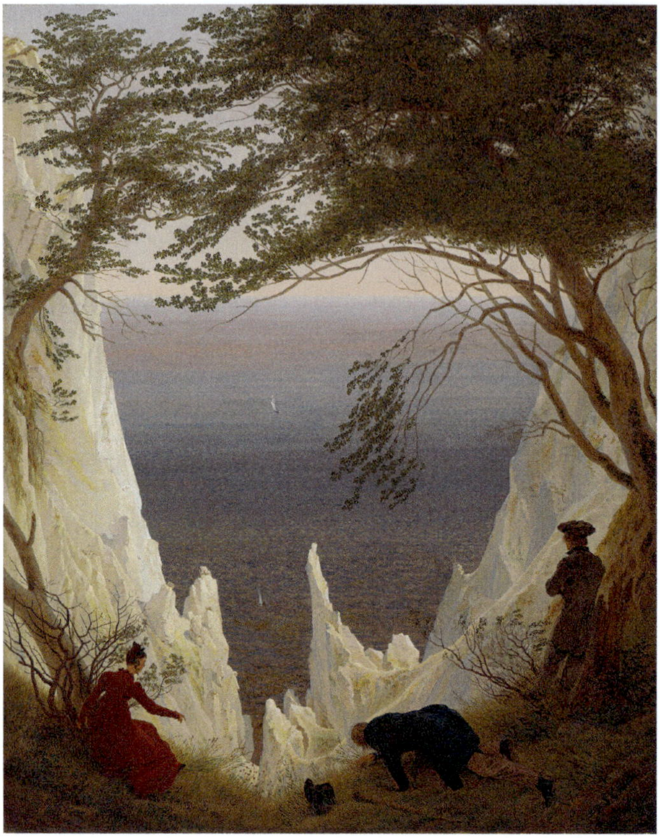

Figure 6.6 Caspar David Friedrich, *Chalk Cliffs on Rügen*, 1819. Oil on canvas, 90 × 70 cm. Winterthur, Museum Oskar Reinhart, inv. no. 165.

Source: SIK-ISEA Zürich (Philipp Hitz).

by the sun.[18] The same combination of warm and cool colours is also evident in the waves, which – in the words Carus used when he was 'inspired' by Friedrich to visit Rügen in 1819 – appear 'gold and violet'.[19] This argument gains credence from Minnaert's observation that an 'unbiased' observer will see distinctly blue shadows on a white surface such as hoar-frost when viewing them from within a 'dense, dark fir wood' because this mode of viewing imitates 'aperture' seeing, which overrides 'constant' local colour.[20]

Just how aperture seeing of the kind enabled by framing a scene makes shadow colour manifest is hinted at in David Katz's important book, *Der Abfau de Farbenwelt* of 1930. This describes how framing a motif with 'pieces of paper' defeats 'memory colour' by making it difficult to recognise the objects in a scene, thereby rendering 'aerial perspective' more noticeable.[21] By extension, other modifications to local colour, including coloured shadows, become more perceptually salient under these circumstances as well. Hence, Katz argued, a similar effect occurs when the painter 'bends over or looks under his arm at the landscape in order not to be influenced by the "genuine" colours of objects' in an effort to 'apprehend the colour value to be placed upon the canvas'.[22]

The psychologist Irwin Priest employed a similar argument in an article of 1906, which may explain how the Impressionists and others enhanced their perception of the coloured shadows that appear on snow. Here, Priest contended that we do not experience the 'saturation' of 'blue shadows' maximally when we are 'fully conscious' of the snow on which they fall and that by corollary this 'increases in a most marked degree' when our awareness of it is 'eliminated' or 'reduced'.[23] Shadows are consequently 'much more saturated', he maintained, in the 'image on a camera ground glass' than they are in a 'direct view of the landscape' because this mode of viewing them removes contextual clues as to the snow's identity.[24] Similarly, Marcel Minnaert stated that 'snow-shadows' appear 'much bluer' than normal when seen within the frame of the camera's 'ground glass plate'.[25]

For his part, Katz recommended viewing objects in a 'plane mirror' which is 'turned through a certain angle from its true position' in order to defeat memory colour.[26] It is significant, then, that Carus discussed the pros and cons of observing the 'colours' of a 'natural landscape in a mirror' in a letter he completed in 1820.[27] It may even be that he took this idea from Valenciennes, who had suggested viewing the landscape in this way in order to render its 'perspective' apparent.[28]

The Pre-Raphaelites' White Card

The painter John Brett, a satellite of the Pre-Raphaelite Brotherhood, used a more radical viewing device in the form of a reduction screen. This is apparent from a letter of 1858 in which he advised his sister, Rosa, that if she wanted to 'paint the things the right colour', she should 'Look at the real ones through a hole in a card as big as a nut to see what they are'.[29]

Brett's procedure can be reconstructed in greater detail from *The Elements of Drawing* of 1857 by John Ruskin, whose writings the artist greatly admired.[30] Here, with the aid of diagram showing how the card should be held so as to catch the light appropriately, Ruskin urged the painter to:

> [P]lace yourself at a window which the sun does not shine in at, commanding some simple piece of landscape . . . then take a piece of white cardboard, cut out a hole in it about the size of a large pea. . . . Then you will see the landscape, bit by bit, through the circular hole. Match the colours of each important bit as nearly as you can . . . beside the aperture.[31]

Crucially for Ruskin, the card disposed its user to see 'flat stains of colour'.[32] Or as he explained elsewhere, it facilitated a 'childish' and 'infantine' way of seeing, amounting to '*innocence of the eye*', in which we 'see nothing but flat colours', 'patches of colour', or 'patches of different colours variously shaded', rather than 'what we only know' or a world of 'solid Form'.[33] Ruskin thus effectively described how the card allowed the artist to see colours pure and simple, rather than colours belonging to the surfaces of things we recognise. Implicitly, then, rather as with the framing devices mentioned earlier, it facilitated the perception of normally elusive shadow colour by allowing the artist to transcend the homogenising effects of memory.

The card, more specifically, allowed the painter to see that the 'ordinary shadows' are 'never black' as Ruskin argued but always have 'some *colour*' as they are 'always of a luminous nature'.[34] Similarly, it would show the painter that 'sunlighted grass' is not merely 'shade and light' but appears 'bluish green barred with gold' to a 'highly accomplished artist'.[35] Ruskin may have had Hunt in mind when he made this remark, as the grass looks like this in *Our English Coasts* (Figure 4.12). A similar effect can be observed in Rosa Brett's *The Artist's Garden* of *c*. 1859–60, particularly at the edges of shadows.[36] The effect of the card can also be discerned in John Brett's painting, *The Stonebreaker* of 1857–58 (Figure 6.7), where intensely blue–violet shadows are apparent, particularly on white surfaces.

Like Ruskin, Katz explained the effect of an 'aperture screen' or 'reduction screen', consisting of a sheet of white or grey card or paper with a hole in it, in terms of its ability to inhibit memory colour.[37] He did so, however, by giving his predecessor's argument a phenomenological twist, declaring that a screen caused colours to appear in 'film' mode as pure colours, rather than in 'surface' mode as the colours of familiar objects.[38] His argument was later expanded by the philosopher, Maurice Merleau-Ponty in his *Phénoménologie de la perception* of 1945. He explained that looking through the 'window of a screen which hides the source of light' frees 'colours from the objectivity they acquire on the surfaces of bodies' and makes them into 'areas of

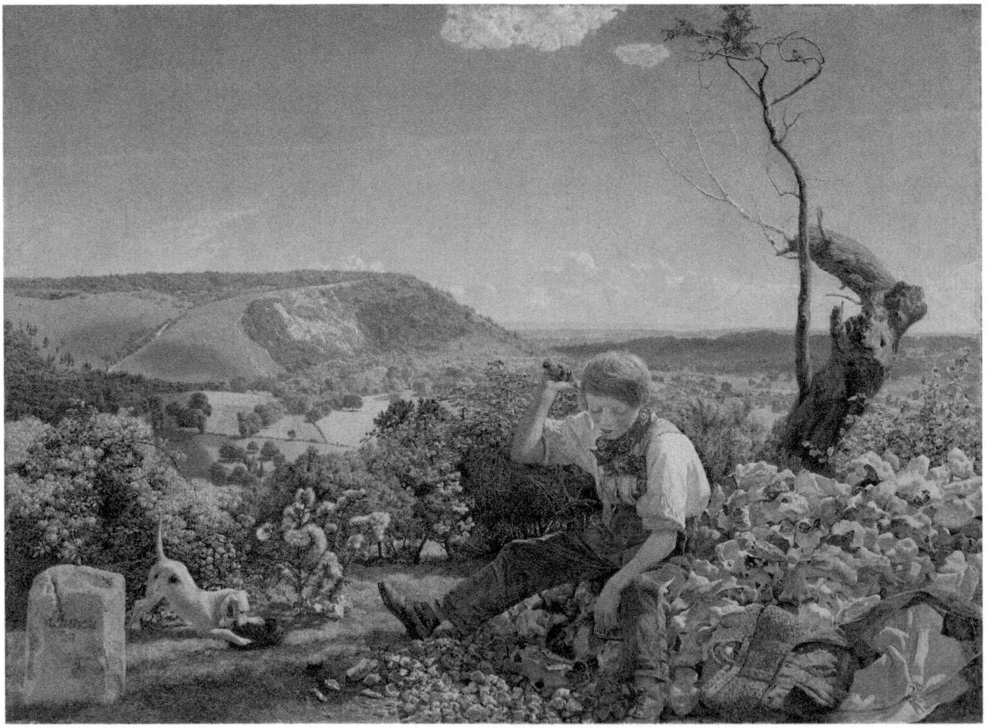

Figure 6.7 John Brett, *The Stonebreaker*, 1857–58. Oil on canvas, 51.3 × 68.5 cm. Liverpool, Walker Art Gallery, National Museums Liverpool, inv. no. WAG 1632.

Source: Walker Art Gallery/Bridgeman Images.

light' and colour instead.[39] Thus, he continued, we 'coloured patches which are all situated on one and the same "fictional" plane', rather than 'real bodies' of a 'determinate colour and having their place in the world' of the kind we recognise.[40]

The arguments of psychologists and philosophers considered so far all imply that frames and reduction screens allow us to see more clearly what we could see under normal circumstances but usually do not, by suppressing the cognitive aspect of constancy. Merleau-Ponty makes a rather different argument, however, in the continuation of his discussion of the effect of a reduction screen. He maintained, more particularly, that the 'screen eliminates' the 'totality of the field' or the wider context of the scene before us, which is 'the decisive factor in the phenomenon of constancy'.[41] By implication, therefore, a screen does not only enhance the perception of shadows simply by inhibiting our knowledge of the surfaces they lie on, it also does so by dampening the tendency of the process of comparison at work in constancy to suppress them in situations where they are not pronounced. It would seem, then, that framing devices sometimes reveal small, objective variations in the colours of areas illuminated by coloured light of the kind Lichtenberg identified.

If so, then, a contradiction at the centre of the viewing practices that such devices facilitate is that they make effects apparent which constancy normally inhibits at the same time as those it creates. Disentangling the two processes concerned is sometimes extremely difficult, moreover. So it is often unclear which kind of shadow a particular practice produces.

Impressionist Squinting and Related Viewing Practices

Like the Pre-Raphaelites, the Impressionists developed a mode of colour-patch vision capable of enhancing shadow colour. But unlike them, they did so without employing devices, simply by half-closing the eyes or squinting. Painters often squinted when beginning a painting in order to see the principal masses of a scene; evidence suggests that the Impressionists did so throughout the painting process. In his 1928 biography of his friend Monet, for example, Georges Clémenceau claimed that the painter looked at the motif with his 'eyes half-closed'.[42] Similarly, Cézanne's close companion Émile Zola described, in his novel *L'Oeuvre* of 1886, how the painter Claude Lantier and his companion Christine stood 'in front of [a] painting for a long time, half-closing their eyes'.[43] And in his novel, *Dans le ciel* of 1892–93, Monet's supporter Octave Mirbeau had the painter Lucien declare that 'when we squint' at a rubbish heap, 'forms appear, shapes of flowers, beings, which burst through the shell of their embryo'.[44]

Joris-Karl Huysmans also suggested as much in his article, 'L'Exposition des Indépendants en 1881', where he remarked that 'Mme Morizot [sic] undoubtedly possesses an eye which, by a special aptitude of the eyelids, squints naturally, allowing her to seize the most tenuous niceties of the stains [of colour] that bodies make on the surrounding air'.[45] To use the language of modern psychology, Huysmans realised that an Impressionist painting suppresses high spatial frequency foveal vision and favours lower spatial frequency vision, which is characteristic of peripheral vision.[46] Huysmans also observed in 'L'Exposition des Indépendants en 1880' that – by corollary – the 'Impressionist system of patches' forced the 'eye to squint in order to re-establish the equilibrium of people and things'.[47] He realised, in other words, that the shapes in a 'blurry' Impressionist painting resolve better when it is viewed at low spatial frequency or at the perceptual scale the painter favoured when squinting.[48] The English critic, Dugald Sutherland MacColl, made similar claims in his article, 'Impressionism', published in the tenth edition of the *Encyclopaedia Britannica* in 1902. Here he described how the Impressionist painting 'surveys the field and determines which of the shapes and tones are of chief importance to the interested eye, enforces these, and sacrifices the rest', with the result that 'minor planes are swamped in bigger' and 'smaller patches of colour are swept up into broader'.[49]

Importantly, both Huysmans and MacColl emphasised how squinting not only eliminates detail but, in so doing, it also allows colour to assume greater perceptual salience. Again, modern science confirms this observation with the idea that the chromatic channel operates at low spatial frequency, which means that colour is most apparent when high spatial frequency detail is suppressed.[50] Merleau-Ponty argued in much the same vein that 'When the painter half-closes his eyes, he does away with the field's organisation in depth and with it, the precise contrasts of lighting, so that there are no longer any determinate things with their own colours'.[51] Thus, the painter who 'isolates' colours 'from their surrounding, by half-closing his eyes' comes to 'see colours as they are determined by the quantity and quality of reflected light' without the aid of a reduction 'screen'.[52] All this strongly implies that Clémenceau was perfectly correct to state that 'while looking at a tree, I see nothing but a tree' but Monet, 'with his eyes half-closed, thinks: "How many tones of how many colours [are there] to the luminous transitions of this simple rod"'?[53]

The Impressionists' habit of squinting led some contemporary critics to suggest that they painted how things appeared in peripheral vision. Among these was the polymath, Georges Guéroult, who argued in an article in 1881 that when the Impressionist painter 'sees a patch', he 'paints a patch' depicting 'objects as they are presented in *indirect vision*, or as they are seen when they are not found sufficiently interesting to be *looked at*'.[54] Guéroult's irony notwithstanding, it is possible that some Impressionists did paint as he suggested. Their colleague, Whistler, for example, was interested in the ideas about peripheral or 'indirect vision' which Helmholtz had articulated in the *Handbuch der physiologischen Optik* and elsewhere.[55] The more specific idea that Impressionist painters sometimes attended to the colours that shadows exhibited in peripheral vision gains credence from an article of 1923 by the psychologist, Adelbert Ames. This argued that 'artists' commonly render 'shadows in out-of-door scenes much bluer than they appear when one looks directly at them' because the 'blue appearance of shadows which are imaged on the side of the retina' is 'very easily seen'.[56]

It is, of course, difficult to remain aware of what lies in peripheral vision. One way of doing so is to cultivate the habit of keeping the eye still by focusing steadily on something else without attending to it, as Cézanne is said to have done.[57] Cézanne did in fact describe a similar mode of viewing when he told Rivière and Schnerb in 1905 that the 'painter's eye' should be 'presumed static'.[58] (It could be that he was alerted to this technique as a result of taking up Valenciennes's recommendation that the painter should view the motif with a 'static eye', by looking at it through a 'window pane' or in a small 'mirror'.)[59] The advantage of viewing a surface this way, Cézanne went on, is that it does not appear 'unified in colour and tone' as it does when 'our eye moves so as to perceive it in its entirety', but instead the 'amount of light' it exhibits 'is not the same for [any] two points' on it.[60] Cézanne seems to have realised, in other words (rather as Goethe had when he noticed the warm colour of a blue shadow's surround), that arresting the eye not only reveals the colours in general favoured by peripheral vision, but also, and more particularly, the small, objective variations in the colour of a surface that are suppressed when the eye scans freely back and forth over it.

The effect of Cézanne's way of looking can be seen in *Apples and Oranges* of *c.* 1899 (Figure 6.8) in which blue shadows feature prominently. It is questionable that these are a function of constancy since they are not discernible in a studio where cool light from a north-facing window is the sole light source. Only a little later, however, Cézanne installed a set of windows in the south-facing wall of the studio at Les Lauves he occupied since 1902, which transmitted a slightly warmer light than the large, north-facing window. It may be, then, that Cézanne used two light sources to illuminate this work. This is also implied by the fact that the work not only depicts cool shadows, but their warm relatives as well. The lone apple at the centre of the painting, for example, casts a spectacular triangular shadow of this kind over the right edge of the plate immediately to its left.

A less drastic, but equally effective, viewing technique artists employed in order to achieve colour-patch vision involved learning to look at the motif with a view to depicting it in discrete touches of colour. Ruskin advocated this way of seeing in *The*

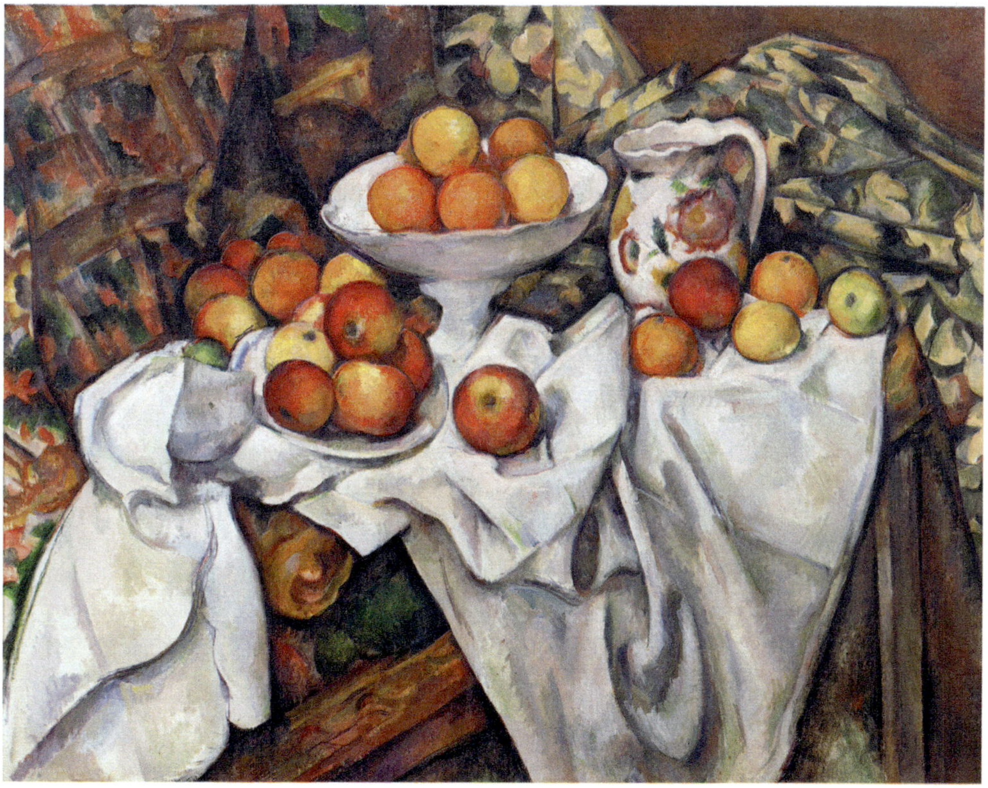

Figure 6.8 Paul Cézanne, *Apples and Oranges*, *c.* 1899. Oil on canvas, 74 × 93 cm. Paris, Musée d'Orsay, inv. no. RF 1872.

Source: Musée d'Orsay/Bridgeman Images.

Elements of Drawing, where he suggested that the painter could 'educate' his 'eye' to perceive 'differences of hue without the help of the cardboard', by laying each touch of a watercolour 'like a mosaic worker', as though 'it were a patch of coloured cloth, cut out, to be fitted neatly by its edge to the next patch'.[61] Ruskin's argument makes sense inasmuch as the intentionality with which we look at something evidently has an effect on what we see it as. His argument also draws support from a body of psychological evidence that looking at an organised visual field affects how we see the world when we subsequently look away from it.[62] It makes sense, in other words, to think that looking at a sketch made out of patches of colour inclined the artist to see the motif in the same terms.

Monet may have read *The Elements of Drawing*.[63] If so, then it is no coincidence that his informal pupil, Lilla Cabot Perry, recalled in 1889 that he advised her: 'When you go out to paint, try to forget what objects you have before you' and 'merely think, here is a little square of blue, here an oblong of pink, here a streak of yellow, and paint it just as it looks to you.'[64] It would seem, in any event, that looking at the world with a view to painting it really did dispose Monet to see shadow colours with greater vividness than normal. This conclusion is supported from a statement the artist made in November 1903 to Count Harry Kessler, who had asked him 'how he came to use colours for the shadows' in his *Women in the Garden* of 1866, and particularly the 'light blue shadows' in the dress of 'the woman at the left'.[65] 'Ah well', Monet told him, 'it was by egging each other on, Renoir, Bazille, and me. One said: Look, how wonderful it is, this tone here, that tone there. You should paint that, it would be fabulous'.[66]

Perry also recalls that Monet told her that he 'wished he had been born blind and then had suddenly gained his sight so that he could begin to paint' without 'knowing what the objects were that he saw before him'.[67] Again, this statement could indicate the painter's interest in Ruskin, who had argued in *The Elements of Drawing* that to see patches of colour 'merely as such' was to see 'as a blind man would see if suddenly gifted with sight'.[68] Monet's words also suggest he may have read Hippolyte Taine's *De l'Intelligence* of 1870, which discussed how people 'born blind' see 'patches' of colour immediately after gaining sight.[69] Taine even argued that 'Painters in colour' were 'aware' of the 'state' of vision experienced by the blind man and were prone to 'revert to it' and hence to 'seeing their model as a patch' consisting only of 'colour'.[70]

Ruskin's and Taine's claims about the sight of those recovering from blindness are questionable, just as their claims about naivety are.[71] But rather as misleading science sometimes enhanced earlier artists' perception, the ideology of naivety had a virtuous result of its own. That is, the way of seeing the Impressionists cultivated was one which Katz later described as 'dominant in the painter', namely a 'subjective attitude' which 'causes colours to become more similar to the colours appearing after reduction' or to supress their tendency to appear as 'memory colours'.[72] Irrespective of the reasons behind it, therefore, the subjective attitude nurtured by the Impressionists really could have enhanced their perception of shadow colour.

Notes

1. On the blueness of Blechen's shadows, see the catalogue entry by Stephano Tumidei in Cavina 2001, 294. For an extended description of Blechen's coloured shadows, see Schöller 2017, 40–49. On Blechen, see also Busch 1994, 281–282.
2. Carus 2002, 85.

3. See Wittgenstein 1995, 198e. On the painting as a criterion in this sense, see Wollheim 1973, 5. On Wittgenstein and the criterion more generally, see Garver 1994, 177–196.
4. Carus 2002, 85.
5. As do earlier paintings by Friedrich and Johan Christian Dahl who moved to Dresden where Friedrich was working in 1818, after which they lived in the same building from 1823. See Vaughan et al. 1972, 38–39.
6. See Butlin 2005, 35 and for a transcription, Lister 1988, 34–35.
7. Butlin 2005, 35 and Lister 1988, 35.
8. Butlin 2005, 193–194 and Lister 1988, 46.
9. Butlin 2005, 195 and Lister 1988, 46.
10. Hunt also used this quick-drying resin medium, which greatly facilitated oil sketching by making it possible to deposit a quantity of paint in one application without it running. See Smith 2012, 19 (citing a communication from Joyce Townsend). Brown mentions his use of copal in Brown 1981, 6, 7, 9, 35, 48–50, 52–53, 64, 70, 76, and 183. On Hunt's use of copal, drying oil, and tube paint, see Katz 1995, 161.
11. Brown 1981, 145. Cited in Staley 1973, 39 and Treuherz et al. 2011, 170. See also Brown 1981, 90 for a diary entry of 5 September 1854, which describes how he set out, 'About 3 out to a field, to begin the outline of a small landscape. Found it of surpassing loveliness. Corn shocks in long perspective form, hayricks, and steeple seen between them foreground of turnips blue sky and afternoon sun. By the time I had drawn-in the outline they had carted half my wheat: [B]y to-day all I had drawn in was gone'. Cited in Staley 1973, 40.
12. See Wollheim 1973, 5–9.
13. See Brown 1981, 90, on how it began as a painting of 'blue sky & afternoon sun' on a Monday (3) in early September 1854; and ibid., 131, which records that the artist was still working on the painting on 7 April 1855.
14. Valenciennes 1799, 147. For a putative connection between Friedrich's work and Valenciennes's text, which was translated into German in 1803, see Grave 2012, 173.
15. The perceptual issues are addressed in Helmholtz 1891, 94–109 and Hardin 1988, 25.
16. See Carus 2002, 3 and 53, n. 16.
17. On Friedtish's visit to Rügen, see Vaughan et al. 1972, 111. For a discussion of the relation between Friedrich's finished paintings, imagination, and the studies made before nature, see Amstutz 2020, 151.
18. See John 2005, 7, for the idea that these (and the 'grey' tones alongside them) either are a function of the 'weather' or represent 'impurities' in the rock.
19. Carus 1941, 5 and 19.
20. Minnaert 1954, 136–137. The reduction screen is to be sharply distinguished from the reduction tube, which makes the shadow colour disappear precisely by isolating it from the rest of the visual field. It follows that Minnaert is incorrect to suggest here that the experience he describes is tantamount to viewing the scene through a 'tube with an opening at the end'.
21. Katz 1935, 161 and ibid. 162, which notes that when the frame is taken away, the memory colour comes back 'into play'. See also Hering 1964, 7–8, for the argument that we 'imagine snow as white', as with all objects, because we normally see its '"real" colour', 'beyond and through' any 'shadow' falling over it, in terms of the 'memory colour' we have built up for it over time.
22. Katz 1935, 171.
23. Priest 1926, 308. See also Churma 1994, 4722, which argues that the effect Priest describes depends on the fact that we do not see the shadow (rather than the snow) as such.
24. Priest 1926, 308.
25. Minnaert 1954, 137.
26. Katz 1935, 171.
27. Carus 2002, 91.
28. Valenciennes 1799, 131.
29. Cited in Bennett 1988, 18; Brett 2005, 615; and Payne 2012, 35. On this card, see Hilton 1970, 136. See also Smith 2004, 17, for the argument that Brett used it to see 'absolute' (or isolated) colours as opposed to 'apparent' (or contextual) colours as defined by Sweetlove.

30. See Hickox 1996, 521, for indications of Brett's admiration for Ruskin in his diaries from 1851.
31. Ruskin 1857, 210–211. See Brett 2005, 165, which compares this paragraph with Brett's letter. See also Ruskin 1857, 52, for another passage concerning the white card.
32. Ibid., 52.
33. Ruskin 1857, 5–7.
34. Ibid., 231.
35. Ibid., 7. Cited in Hilton 1970, 136 and Kemp 1990, 305.
36. For an illustration of his work, see Staley et al. 2004, 50.
37. See Katz 1935, 26, 55–56 and 71–72, 94, on the film colours generated by looking through an 'aperture screen' or 'reduction screen'. See also Wurmfield 1998, 187–188, which compares Ruskin's card to Katz's 'reduction screen'.
38. Katz 1935, 55–56. On this distinction, see Katz 1935, 7–28.
39. Ibid., 307. See also ibid., 226–227 and 308.
40. Ibid., 307; citing Gelb 1929, 600.
41. Ibid., 308. On this phenomenon, see also Katz 1935, 78–82 and 280 (citing Von Kries); Land 1977, 118–120; and Mollon 2006, 299.
42. Clémenceau 1928, 19.
43. Zola 1886, 140.
44. Mirbeau 2003, 97. Cited in Tedeschi 2012, 174.
45. Huysmans 1883, 252.
46. Spatial frequency is defined in terms of the number of lines per unit area on a sinusoidal grating. We can distinguish many lines at high frequencies and increasingly fewer at lower frequencies. On the low spatial frequency of peripheral vision and how artists exploit this, see Livingstone 2002, 68–83. See also Mather 2013, 23–24, for the suggestion that Monet's colour-patch paintings of Rouen Cathedral exhibit the low spatial resolution characteristic of short-sight.
47. See Wolff 1875, which argued that an Impressionist painting is similar to one 'which must be looked at from fifteen paces while half-closing the eyes'. Cited in White 1984, 55.
48. On the ability of the human visual system to extract facial shapes from 'blurred' or low spatial frequency information, see Harmon and Julez 1973; Harmon 1973; Schyns and Aude 1999; and Livingstone 2002, 71–73.
49. MacColl 1902–03, 414. Cited in Dewhurst 1904, 109. See also MacColl 1902, 10–16.
50. See Livingstone 2002, 165–169 and Mather 2013, 22–24.
51. Merleau-Ponty 1996, 308.
52. Ibid., 307.
53. Clémenceau 1928, 19.
54. Guéroult 1881, 88–89.
55. See Helmholtz 1962, 2: 39–42. See also Gruetzner 2007, 18–19, for Whistler's response to Helmholtz's ideas about 'indirect vision' in his essay (published in English in 1873), 'The Eye as an Optical Instrument'.
56. Ames et al. 1923, 34. For a review of research on the effect of retinal eccentricity on colour perception, see Hansen et al. 2009.
57. See Ehrenzweig 1953, 196–197. See also Katz 1935, 168 for the argument that it is possible to 'see reductively', 'without a reduction-screen' simply by employing 'eye-movements' which are 'different' from those employed in 'normal vision'.
58. Rivière and Schnerb 1978, 88.
59. Valenciennes 1799, 131.
60. Ibid. Cézanne added that the 'painter' who represents this surface with a 'monochrome layer' of paint does so 'untruthfully'.
61. Ruskin 1857, 213.
62. On the effect that fixating on a pattern has on what is seen immediately afterwards, see McCollough 1965 and Blakemore and Sutton 1969.
63. See Dewhurst 1911, 295, for Monet's assertion that 'ninety percent of Impressionist theory' could be found in Ruskin's *Elements of Drawing*. See also Stuckey 1984, 119, n. 4.
64. Perry 1927, 120.
65. Kessler 2011, 310. For a description of the shadows in this painting, see Schöller 2017, 91–92.

66. Ibid. See also Kessler 2017, for a diary entry from 25 November 1903 which recounts Georges Viau's recollection that Renoir had told him 'how he and Monet came to paint coloured shadows (blue) instead of grey' because 'one day, at Louveciennes, they were painting outside, in the fields, and he finds he has no more black, so he took blue instead. In the evening, they were amazed at the beautiful effect it produced'.
67. Perry 1927, 120. See also Borély 1926, 493, for the statement Cézanne made in 1902 that he wanted to 'See like a newborn!' and Gasquet 1921, 83, for a statement of around 1900 in which he claimed he saw 'in patches'. See also Roux 2007, 21–22, n. 48 for a summary of Impressionist statements about colour-patch vision.
68. Ruskin 1857, 6.
69. Taine 1870, 2: 121–122. Cited in Stuckey 1984, 108–109. On Taine's theory, see Shiff 1991, 173, n. 23 and Smith 2013, 105–106.
70. Taine 1870, 2: 122. On Taine and the 'stain' (*tache*), see also Sjåstad 2014, 47–48.
71. For an overview of recent research into the recovery of sight among people born blind, see Fine 2010.
72. Katz 1935, 167.

7 Aesthetic Value

The Most Beautiful Blue

There would be little point in artists working so hard to capture the phenomenology of coloured shadows if this had no aesthetic value. Fortunately, as already indicated, this is not the case. Indeed, Rumford not only claimed that the shadow cast by the candle in his famous experiment was 'the most beautiful blue that it is possible to imagine', but also that he had 'never found any body to whom I showed these experiments whose eyes were not fascinated with their bewitching beauties'.[1] He as good as claimed, in other words, that the beauty of coloured shadows was a matter of universal assent.

By way of explanation, Rumford suggested three possibilities. The first depended on the 'striking brilliancy and beauty' of shadow colour.[2] What Rumford hit on here is how the blue of a coloured shadow is saturated and luminous at the same time, rather like the blue of the sky, and unlike the blue colours we see on surfaces, which are dark when saturated. Put differently, there is an incongruity to coloured shadows, inasmuch as they have a celestial quality even though they lie on terrestrial objects. They are arguably objects of wonder on this account, as artists did not fail to observe, since they challenge the conceptual categories the spectator brings to them and invite a pleasurable 'free play' between the mind's cognitive capacities and visual experience.[3]

The second reason Rumford gave for thinking coloured shadows are beautiful was that the 'most perfect harmony' existed between the pairs of 'colours' they exhibited.[4] He even declared that this was 'as perfect and pleasing when the shadows were of different tints of brown, as when one of them was blue and the other yellow'.[5] Although the last combination is frequently encountered, brown shadows are rarely seen. Monet nevertheless recorded shadows of a muddy green bordering on brown in *Stack of Wheat (Snow Effect, Overcast Day)* of 1890–91 (Figure 7.1), where they harmonise perfectly with the muted red tones in the stack.

Rumford's third argument was that a 'great part of the pleasure' afforded by coloured shadows 'arose from the continual changes of colour, tint, and shade, with which the eye was amused, and the attention kept awake' by them.[6] He may have had in mind here how shadows change colour as the filter colouring the artificial light source illuminating them is switched. Something similar happens in nature as the sun sinks or rises, although it is difficult for artists to capture such changes as they occur. When seen together, however, this process is apparent in Monet's series paintings.

The rider Rumford added to this last observation suggests he may have meant something else, however. This stated that whereas with objects, 'We are used to seeing

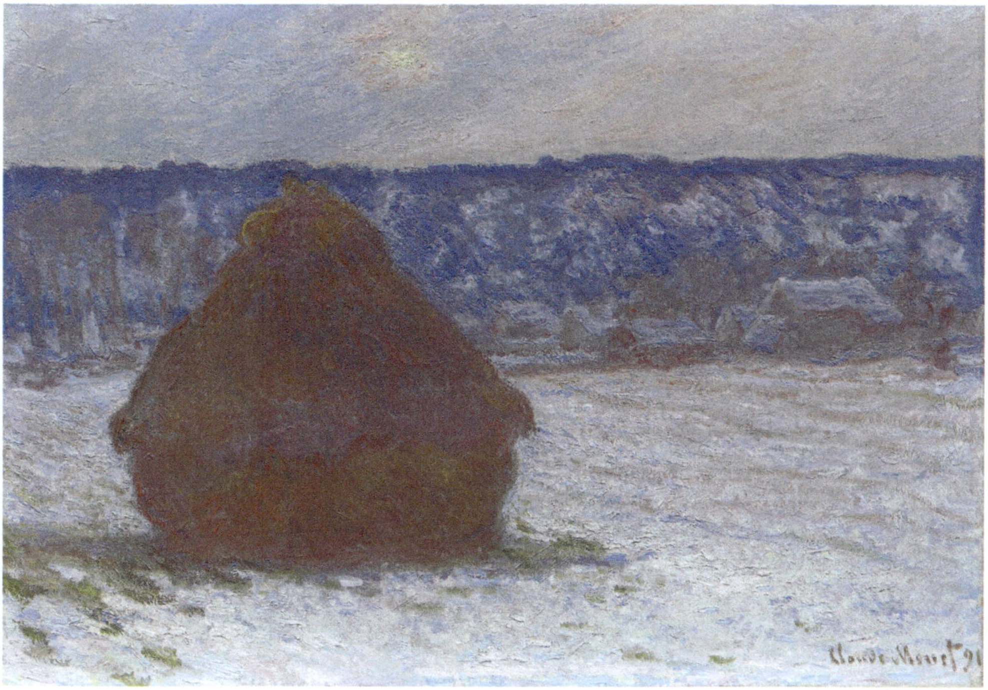

Figure 7.1 Claude Monet, *Stack of Wheat (Snow Effect, Overcast Day)*, 1890–91. Oil on canvas, 66 × 93 cm. Chicago, The Art Institute of Chicago, Mr. and Mrs. Martin A. Ryerson Collection, ref. no, 1933.1155.

Source: The Art Institute of Chicago/Bridgeman Images.

colours fixed and unalterable', 'motionless', and 'consequently dead, and tiresome to the eye', with coloured shadows, 'all is motion, life, and beauty'.[7] What this implies is that shadow colour is not stable but continually shifts, perhaps in response to fluctuations in the colour of an artificial light source or variations in natural light. Again, it is difficult for artists to show how this process unfolds in a static painting. The sense that shadow colour is changing is nevertheless indicated by some of Monet's grain stacks, perhaps because his blurry brushstroke is conducive to the impression of movement and hence to the sense that time is flowing.[8]

Pure Consciousness

The most fertile aspect of Rumford's argument, however, is implicit in its somewhat obtuse claim that the colour of shadows is not as 'hard as the solid as the solid bodies' casting them.[9] What Rumford seems to have meant is that shadow colour, unlike the surface colour of objects, has a peculiar spatial indeterminacy which interferes with, rather than facilitates, our perception of solidity.[10] Certainly, by dint of appearing in 'film' mode, they appear to lie in a fictional plane of the kind mentioned by Merleau-Ponty.[11] Or as Katz said of film colours, they have a 'frontal-parallel orientation' and appear spatially 'homeless' and 'spongy' like the colours of the sky or fog.[12]

The aesthetic implication of this behaviour is that shadow colour defeats what Merleau-Ponty describes as our inclination to respond to the tangibility of objects by reaching out and grasping them. The basis of this contention is the claim (he made in 'Le doute de Cézanne' of 1945) that the 'distinctions between touch and sight are unknown within primordial perception' or the primitive substrate of vision.[13] (And of late, this argument has been borne out by the discovery of a 'vision for action' system in the brain, which uses visual input to direct unconscious motor responses, alongside the more familiar 'vision for perception' system, which gives rise to conscious visual experience.)[14] Hence, as Merleau-Ponty argued in *L'Oeil et l'esprit* of 1960, colour does not simply give shape to 'the visible' by showing how objects come into being in 'depth'.[15] Rather, every 'colour' also has a 'definite motor value', whereby it elicits a physical response such as grasping or withdrawing the hand.[16] There is a sense, then, in which the coloured surfaces we see get under our skin. Or to use Merleau-Ponty's metaphor in *Le visible et l'invisible*, seeing involves a 'chiasm' wherein our 'flesh' and that of the world cross over into one another.[17] It follows, by corollary, that in supplanting palpable surfaces in our awareness, coloured shadows defeat this kind of reciprocity. Seeing coloured shadows is thus a highly distinctive experience because it involves an imaginatively 'disembodied' way of seeing, which releases us from our normal attachment to the world.

Granting this much, seeing of this kind has an ethical value because it loosens the normally ineluctable constraint on 'freedom' that our routinely 'fleshly' vision imposes by making 'our bonds with the world' unbreakable.[18] That is, they offer a glimpse of what it might be like if we had 'pure consciousness' and hence were 'capable of anything'.[19] It could be objected that fireworks, neon signs, and the like have the same merit, but this would miss the point, which is that coloured shadows have poignancy because they fall upon objects which are real and present, and hence defeat the responses such things characteristically elicit. Seeing coloured shadows is thus actively, or dynamically, liberating.

Notes

1. Thompson 1794, 116–117.
2. 1794, 116. See also Buffon 1746, 158, which mentions the 'very lively blue' of a coloured shadow.
3. See Fisher 2003, 39, 55, and 131. See also Kant 2007, 48–49.
4. Thompson 1794, 117.
5. Ibid.
6. Ibid.
7. Ibid.
8. See Mather 2013, 24, on Monet's 'blurred' vision; and ibid., 108 on how 'the motion blur visible in photographs can be employed in paintings to convey a sense of movement'. On the importance of blur as 'kinetic texture' in a painting, see Braddick 1995, 209–213. On how blur mimics effects of 'shearing', 'wiping', and 'unwiping' that the edges of forms produce in normal perception when moving across a background, see Bruce et al. 1996, p. 262. See also Vollard 1936, 156, for Manet's claim that: 'There can be no question of wire-drawn lines in a moving thing' such as the Venetian lagoon, 'only of values which, rightly observed, will constitute the real volume, the unquestionable design'. On Monet's 'blurred' vision, see Mather 2013, 24. For the claim that Monet's work produces the 'illusion of movement' by using 'equiluminant' colours, see Livingstone 2002, 153 and 157.
9. Thompson 1794, 117.

10. See Prettejohn 2000, 163, which makes the similar claim that William Holman Hunt used 'vivid coloured shadows' to 'characterise the shadow as a thing in itself' and not as 'a foil to the brighter areas of the picture'. See also Hering 1964, 8, for the claim that normal, 'finely graded shadows on the surface of a body' serve to 'mediate the perception of its form, its relief, and its distance', because they are easily 'separated as something accidental from the colour of the surface' beyond them.
11. Merleau-Ponty 1996, 307.
12. Ibid., 8, 11, and 71–74.
13. See Johnson 1993, 65.
14. On the support science lends to phenomenology, see Carman 2008, 111–112. For a vision for action, see Milner and Goodale 1995; O'Regan and Noë 2001; and Jacob and Jeannerod 2003.
15. Merleau-Ponty 1993, 141–142.
16. Merleau-Ponty 1996, 209–210. See also Hardin 1988, 150–151 and Smith 2014, 93–97, 129–133, and 137–138.
17. See Merleau-Ponty 1968, 130–155.
18. Merleau-Ponty 1996, 440. See also Johnson 1993, 74.
19. Merleau-Ponty 1996, 440.

8 Conclusion

In his translation of Thomas à Kempis's *The Imitation of Christ*, published in 1653, Corneille elaborated on its author's warning that 'not every desire' that appears to derive 'from the Holy Spirit' really does so by saying that some possessed only a 'coloured shadow' of 'virtue'.[1] Plainly enough, if Corneille meant to refer to coloured shadows of the kind this book is concerned with, he thought they were not only insubstantial, but downright untrustworthy, too.

I have tried to argue something different. I have suggested that these marginal phenomena are fascinating not least because for several hundred years they posed a challenge to science to provide an explanation. More importantly perhaps, although Monge's theory of constancy was perfectly capable of doing this, no straightforward 'paradigm shift' occurred at the end of the eighteenth century whereby it trumped the theory of contrast any more than this displaced the theory of reflection earlier in the century.

There is no obvious explanation for these facts, although it is likely that the simplicity of the earlier forms of explanation gave them an unfair advantage. The processes of reflection and contrast are in any event easier to picture than some of those involved by constancy. Reflection is readily envisaged – as Leonardo did – as a process in which colour leaves the surface of one object with the light it reflects and arrives with it in the shadow. Similarly, as Josef Albers described, the effect of contrast can be pictured in terms of how the surround or 'ground' 'subtracts its own hue' from the colour it encloses and vice versa.[2] By analogy, adaptation can be imagined as what happens when the colour is removed from the dominant illumination; but the computational processes involved by constancy and their neural implementation are much more elusive and perhaps even 'unpicturable' in principle, rather, as sub-atomic particles are by some accounts.[3]

The case of coloured shadows also sheds light on the claim that seeing amounts to, or involves, seeing as and seeing that. The verbal reports of scientists and artists from Leonardo onwards leave little doubt that theories did define how they conceived of the processes responsible for generating coloured shadows. And it would seem, too, that theories shaped what observers of both kinds saw. To this extent, they distorted artists' perceptions by inclining them to see these phenomena as the outcome of causes which they could not in fact generate on their own.

Notwithstanding, theories did not prevent observers who relied on direct experience from seeing coloured shadows in their full glory. Artists even enhanced their perception of coloured shadows either indirectly by developing devices which served this purpose or through the act of painting or directly by looking at them with the appropriate intentionality. It would seem, then, that the lure of their phenomenology was stronger than the attraction (or distraction) of concepts for most artists. This is not to say, however, that painting succeeded where science failed, merely that it succeeded in the domain of description as opposed to explanation.

Last, but not least, coloured shadows are eminently worth looking at – if Merleau-Ponty is to be believed – because they allow us to detach ourselves imaginatively from the world. Lichtenberg seems to have alluded to this ability metaphorically, when he told Goethe in his letter of 1793 that he was so enthused by 'coloured shadows' that he had been pursuing them 'like a little boy chasing butterflies'.[4] And for his part, Goethe compared the scene created by coloured shadows on the Brocken, in which 'every object had clothed itself in two vivid and so beautifully harmonising colours', to a 'fairy world'.[5] Goethe expressed a similar sentiment in 'Von den farbigen Schatten', where he argued that the 'landscape painter' can only attain a 'high level in his art' when he 'creates a magical world by combining these celestial phenomena with the shapes and colours of earthly objects'.[6] And arguably, this is exactly what Runge achieved in *Morning*, where coloured shadows are so extensive that they create an ethereal alternative to the tangible world in which hovering figures enjoy freedom from earthly constraints.

The dematerialising power of coloured shadows was not lost on Pissarro, even if he was unconcerned with its existential dimension. As I have described elsewhere, he made a number of paintings of the early 1880s, including *Apple Picking* of 1881 (Figure C.1), in which women wearing blue garments are caught up in and partly

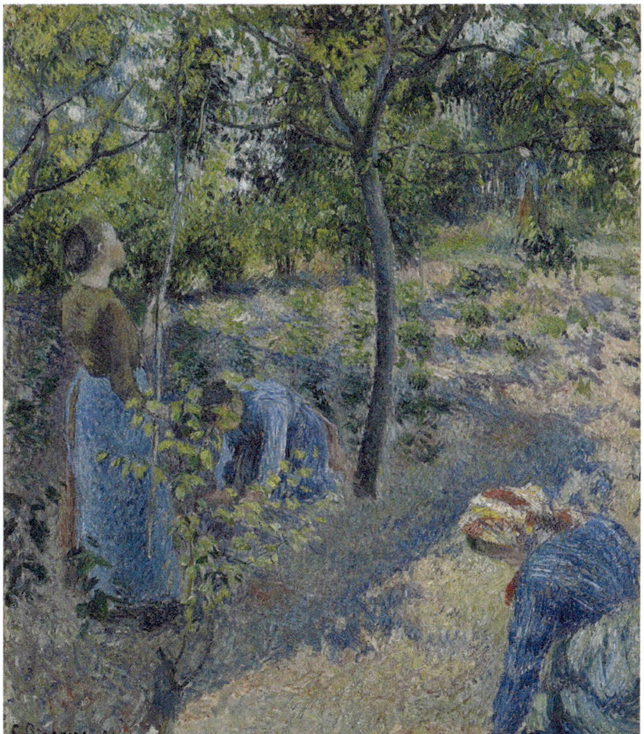

Figure C.1 Camille Pissarro, *Apple Picking*, 1881. Oil on canvas, 64.8 × 54.3 cm. Private collection.

Source: © Christie's Images/Bridgeman Images.

dissolved by extensive blue shadows.[7] The significance of these works is related to Pissarro's anarchist beliefs and demotic sympathies. For instance, his preference for blue as the colour of the workers is echoed in the opinions articulated by the Impressionist painter, Vibrac, in Paul Adam's novel *Soi* of 1886, who not only painted 'mauve or blue shadows', but also admired how 'The blouse of very poor workers' was a 'dead blue' which engendered 'extraordinary, greenish shadows'.[8] In addition, Pissarro expressed a conviction in his letters that painting immaterial effects of light and colour provided an alternative to the kind of 'muddy' painting he identified as 'bourgeois', which was solely concerned for the materiality of things.[9] The extensive blue shadows in his paintings thus instantiate a populist and dematerialised alternative to the reified bourgeois norm. And even if naively utopian in this regard, they nonetheless succeed in conveying how coloured shadows can offer a measure of freedom from the constraints imposed by the lure of things. And this – *pace* Kempis – is a substantial virtue.

Notes

1. Kempis 1654, 88.
2. Albers 2013, 20.
3. See Hanson 1963, 42–47.
4. Joost et al. 2002, 303.
5. Goethe 1840, 35.
6. Goethe 1897, 123–124.
7. See Smith 1992.
8. Adam 1886, 221 and 416.
9. Ibid., 227.

Select Bibliography

Works Published up to 1900

Adam, Paul. *Soi*. Paris: Tresse & Stock, 1886.

Andersen, Hans Christian. *The Improvisatore: Or, Life in Italy*. Translated by Mary Howitt. 2 vols. London: Richard Bentley, 1845.

Anon. 'Annales de Chimie, & c. Chemical Annals, or a Collection of Memoirs Relative to Chemistry and to the Arts Depending on It'. Appendix, *The Monthly Review or, Literary Journal Enlarged* 11 (May–August 1793), 515–544.

Anon. 'Varieties, Literary & Philosophical With Notices Respecting Men of Letters, Artists, and Works in Hand'. *The Universal Magazine of Knowledge and Pleasure* 17, no. 103 (June 1812), 486–495.

Anon. *Catalogue of Pictures by Rubens, Rembrandt, Vandyke, and Other Artists of the Flemish and Dutch Schools*. London: British Institution, 1815.

Anon. [Achille Piron], and Eugène Delacroix. *Eugène Delacroix: sa vie et ses oeuvres*. Paris: Jules Claye, 1865.

Barrow, Isaac. *Lectiones XVIII*. London: William Godbid, 1669.

Barrow, Isaac. *Lectiones opticae & geometricae*. London: William Godbid, 1674.

Beguelin, Nicolas de. 'Sur les ombres colorées'. In *Mémoires de l'Académie Royale de Prusse: année MDCCLXVII*, 27–40. Berlin: Haude et Spener, 1769.

Benson, William. *Manual of the Science of Colour*. London: Chapman & Hall, 1871.

Benson, William. Letter of 7 October 1895 to an Unknown Correspondent. 2ff. Manuscript, Author's Collection.

Blanc, Charles. *Grammaire des arts du dessin*. Paris: Renouard, 1867.

Bossi, Giuseppe. 'Saggio di richerche intorno all'armonia cromatica'. *Memorie dell'Imperiale Regio Istituto del Lombardo-Veneto* 2 (1821), 277–316.

Bouguer, Pierre. *Traite d'optique sur la gradation de la lumière*. Paris: Guerin & Delatour, 1760.

Bourgeois, Charles. 'Sur un nouveau phénomène d'optique'. *Bulletin des sciences mathématiques, astronomiques, physiques et chimiques* 9, no. 3 (March 1828), 179–184.

Brewster, David. *A Treatise on Optics*. London: Longman, Rees, Orme, Browne & Green and John Taylor, 1831.

Brisson, Barnabé. *Notice historique sur Gaspard Monge*. Paris: Plancher, 1818.

Brougham, Henry. 'Experiments and Observations on the Inflection, Reflection, and Colours of Light'. *Philosophical Transactions of the Royal Society of London* 86 (1796), 227–277.

Brown, Ford Madox. *The Exhibition of Work and Other Paintings by F. M. Brown at the Gallery, 191 Picadilly*. London: M'Corquodale & Co., 1865.

Brücke, Ernst. *Des couleurs au point de vue physique, physiologique, artistique et industriel*. Paris: Bailliere, 1866.

Brücke, Ernst, and Hermann von Helmholtz. *Principes scientifiques des beaux-arts: essais et fragments de théorie, suivi de l'optique et la peinture*. Paris: Germer Baillière, 1878.

Burtin, Xavier de. *Traité théorique et pratique des connoissances qui sont nécessaires à tout amateur de tableaux*. 2 vols. Brussels: Weissenbruch, 1808.

Callcott, Augustus Wall. Abstract of Thompson 1799, 1810–20. Manuscript. Oxford, Bodleian Library, Ms Eng. c. 2728, ff36r–37r.

Calvi, Girolamo. *Della norma che per dipingere le ombre*. Milan: Giacomo Piroloa, 1842.

Carus, Carl Gustav. *Mnemosyne: Blätter aus Gedenk- und Tagebüchern*. Pforzheim: Flammer und Hoffmann, 1848.

Castel, Louis-Bertrand. *L'optique des couleurs*. Paris: Briasson, 1740.

Cennini, Cennino. *Trattato della pittura*. Rome: Paolo Sabriucci, 1821.

Chaptal, J.A. *Elémens de chimie*. 3 vols. Montpellier: Jean-François Picot, 1790.

Chase, John. *A Practical Treatise on Landscape Painting and Sketching from Nature in Water Colours*, edited by James Harris. London: J. Barnard, 1861.

Chevreul, Michel-Eugène. 'Mémoire sur l'influence que deux couleurs peuvent avoir l'une sur l'autre quand on les voit simultanément'. *Mémoires de l'Académie Royale des Sciences de l'Institut de France* 11 (1832), 447–520.

Chevreul, Michel-Eugène. *De la loi du contraste simultané des couleurs*. Paris: Pitois-Levrault, 1839.

Chevreul, Michel-Eugène. 'Physical Investigation on Dyeing'. In *Chemical Reports and Memoirs*, edited by Thomas Graham, 165–238. London: The Cavendish Society, 1848.

Chevreul, Michel-Eugène. *The Principles of Harmony and Contrast of Colours, and Their Application to the Arts*. Translated by Charles Martel [Thomas Delf]. London: Longman, Green, and Longmans, 1854.

Chevreul, Michel-Eugène. *The Laws of Contrast of Colour*. Translated by John Spanton. London and New York: George Routledge, 1857.

Claretie, Jules. *La vie à Paris*. Paris: V. Havard, 1881.

Clark, John Heaviside. *A Practical Essay on the Art of Colouring and Painting Landscapes in Water Colours*. London: Orme, 1807.

Cloquet, Jean-Baptiste-Antoine. *Nouveau traité élémentaire de perspective à l'usage des artistes et des personnes qui s'occupent du dessin*. Paris: Bachelier, 1823.

Clover, Joseph. Sketchbook c. 1835. Manuscript. Norwich Castle Museum, NWHCM: 1939.141.

Darwin, Erasmus. *The Botanic Garden: A Poem in Two Parts*. 2 vols. London: J. Johnson, 1791.

Darwin, Robert. 'New Experiments on the Ocular Spectra of Light and Colours'. *Philosophical Transactions of the Royal Society of London* 76 (January 1786), 313–348.

da Vinci, Leonardo. 'Libro di pittura'. Manuscript. Vatican City: Biblioteca Apostolica Vaticana, MS Urb. Lat. 1270, c. 1540 148v.

da Vinci, Leonardo. *Traité de la peinture*. Translated by Roland Fréart and Sieur De Chambray. Paris: Jacques Langlois, 1651a.

da Vinci, Leonardo. *Trattato della pittura*, edited by Raphaël Trichet du Fresne. Paris: Giacomo Langlois, 1651b.

da Vinci, Leonardo. *A Treatise of Painting*. London: J. Senex, 1721.

da Vinci, Leonardo. *A Treatise on Painting*. Translated by John Francis Rigaud. London: J.B. Nicholls, 1835.

Davy, Humphry. 'Some Experiments and Observations on the Colours Used in Painting by the Ancients'. *Philosophical Transactions of the Royal Society of London* 105 (1815), 97–124.

Delabarre, Edmund B. 'Colored Shadows'. *The American Journal of Psychology* 2, no. 4 (August 1889), 636–643.

Delacroix, Eugène. *Journal d' Eugène Delacroix*. 3 vols. Paris: Plon, 1893.

Duranty, Louis-Edmond. *La nouvelle peinture*. Paris: Dentu, 1876.

Duret, Théodore. *Les peintres impressionnistes*. Paris: Heyman & Perois, 1878.

Edgeworth, Maria. *Practical Education*. 2 vols. London: J. Johnson, 1798.

Edgeworth, Maria. 'Waste Not, Want Not: Or, Two Strings to Your Bow'. In *The Parent's Assistant, or Stories for Children*, 123–191. 3rd edn. London: J. Johnson, 1800.

Edgeworth, Maria. *Leonora*. 2 vols. London: J. Johnson, 1806.

Edgeworth, Maria. *Harry and Lucy Concluded: Being the Last Part of Early Lessons*. 4 vols. London: Hunter and Baldwin, Cradock, and Joy, 1825.

Field, George. *Chromatics or an Essay on the Analogy of Colours*. Manuscript. London: Winsor & Newton, 1811.

Field, George. *Chromatics or, an Essay on the Analogy and Harmony of Colours*. London: Valpy, 1817.

Field, George. *Chromatography or, a Treatise on Colours and Pigments, and on Their Powers in Painting, & c*. London: Charles Tilt, 1835.

Field, George. *Chromatography or, a Treatise on Colours and Pigments, and on Their Powers in Painting*. 2nd edn. London: Tilt and Bogue, 1841.

Field, George. *Chromatics or, the Analogy, Harmony, and Philosophy of Colours*. 2nd edn. London: David Bogue, 1845.

Field, George. *Rudiments of the Painter's Art*. London: John Weale, 1850.

Fromentin, Eugène. *Une année dans le Sahel*. Paris: Michel Lévy, 1859.

Galton, Samuel. 'Experiments on Colours'. *Monthly Magazine* 8, no. 48 (1 August 1799), 509–513.

Gautier, Jacques-Fabien. *Chroa-Génésie ou génération des couleurs*. 2 vols. Paris: Antoine Boudet, 1750–51.

Gautier, Jacques-Fabien. *Observations sur la l'histoire naturelle, sur la physique, et sur la peinture*. 2 vols. Paris: Delaguette, 1752.

Gehler, Johan Samuel Traugott. 'Schatten, blaue'. In *Physikalisches Wörterbuch*. 2 vols. 823–826. Leipzig: Schwickertschen Verlag, 1790.

Gilchrist, Alexander. *Life of William Blake, 'Pictor Ignotus'*. 2 vols. London and Cambridge: Macmillan, 1863.

Goethe, Johann Wolfgang von. *Zur Farbenlehre*. 3 vols. *Historischer Teil: Materialien zur Geschichte der Farbenlehre*. Tübingen: Cotta, 1810.

Goethe, Johann Wolfgang von. *Goethe's Theory of Colours*, edited by Charles Eastlake. London: John Murray, 1840.

Goethe, Johann Wolfgang von. *Conversations of Goethe*, edited by Johann Peter Eckermann. Translated by John Oxenford. London: Smith and Elder, 1850.

Goethe, Johann Wolfgang von. *Conversations de Goethe pendant les dernières années de sa vie, 1822–1832*, edited by Johann Peter Eckermann. Translated by Émile Delerot. 2 vols. Paris: Charpentier, 1863.

Goethe, Johann Wolfgang von. *Goethes Werke. Herausgegeben im Auftrage der Großherzogin Sophie von Sachsen. IV. Abteilung: Goethes Briefe*. 50 vols. Weimar: Hermann Böhlau, 1887–1912.

Goethe, Johann Wolfgang von. 'Von den Farbigen Schatten'. *Goethes Werke: Herausgegeben im Auftrage der Grossherzogin Sophie von Sachsen: II. Abtheilung: Goethes Naturwissenschaftliche Schriften: 5. Band: Erste Abtheilung: Chromatik*, 101–125. Weimar: Hermann Böhlau, 1897.

Guericke, Otto von. *Experimenta nova Magdeburgica de vacuo spatio*. Amsterdam: Johannes Jansson Waesberge, 1672.

Guéroult, Georges. 'Du rôle du mouvement des yeux dans les émotions esthétiques'. *Gazette des beaux-arts* (2nd Period) 23–24 (June–July 1881), 536–542 and 82–90.

Hagedorn, Christian Ludwig von. *Réflexions sur la peinture*. Translated by Michael Huber. 2 vols. Leipzig: Gaspar Fritsch, 1775.

Harris, Moses. *The Natural System of Colours*. London: Laidler, c. 1770.

Hassenfratz, Jean Henri. 'Premier mémoire sur les ombres colorées'. *Journal de l'École Polytechnique* 4, cahier 11 (Messidor An X [1802]), 272–283.

Haüy, René Just. *Traité élémentaire de physique*. 2 vols. Paris: Courcier, 1806.

Hay, David Ramsay. *On the Laws of Harmonious Colouring Adapted to House Painting*. Edinburgh: Lizars, 1828.

Helmholtz, Hermann von. 'On the Relation of Optics to Painting'. In *Popular Scientific Lectures: Second Series*, 73–138. New edn. London: Longmans, Green & Co., 1891.
H.F.T. *Observations sur les ombres colorées*. Paris: Veuve Duchesne and Brussels: Dujardin, 1782.
Hire, Philippe de la. *Mémoires de mathématique et de physique*. Paris: Imprimerie Royale, 1694.
Howard, Frank. *Colour, as a Means of Art*. London: Joseph Thomas, 1838.
Hueffer, Ford Madox. *Ford Madox Brown*. London, New York, and Bombay: Longmans, Green & Co., 1896.
Hummel, Johann Erdmann. *Die freie Perspektive*. 2 vols. Berlin: Friedrich August Herbig, 1825.
Hunt, William Holman. 'The Pre-Raphaelite Brotherhood: A Fight for Art (Part I)'. *Contemporary Review* 49 (April 1886), 471–488.
Huysmans, Joris-Karl. *L'Art moderne*. Paris: Charpentier, 1883.
Jombert, Charles-Antoine. *Méthode pour appprendre le dessein*. Paris: Jombert, 1755.
Kempis, Thomas à. *L'Imitation de Jesus Christ. Livre troisième*. Translated by Pierre Corneille. Paris: Ballard, 1654.
Kopisch, August. 'Entdeckung der blauen Grotte auf der Insel Capri'. In *Italia*, edited by Alfred Reumont, 155–201. Berlin: Alexander Duncker, 1838.
La Farge, John. *Considerations on Painting*. New York: Macmillan, 1896.
Lairesse, Gerard de. *A Treatise on the Art of Painting, in All Its Branches*. 2 vols. London: Edward Orme, 1738.
Laugel, Auguste. *L'Optique et les arts*. Paris: Germer Baillère, 1869.
Laurent, Paul. *Théorie de la peinture. Second partie. Traité de perspective aérienne*. Paris: No Publisher and Nancy: The Author, 1828.
Le Blon, Jakob Christoph. *Coloritto or the Harmony of Colouring in Painting*. London: J.C. Le Blon, 1725.
Leclerc, Georges-Louis, Comte de Buffon. 'Dissertation sur les couleurs accidentelles'. In *Histoire de l'Académie Royale des Sciences*. Année M. DCCXLIII, 147–158. Paris: Imprimerie Royale, 1746.
Leclerc, Georges-Louis, Comte de Buffon. 'Observations sur les couleurs accidentelles, & sur les ombres colorées'. In *Oeuvres complètes de M. Le C.te de Buffon. Tome septième. Histoire naturelle, générale et particulière*. Supplément, Tome Second, 309–340. Paris: Imprimerie Royale, 1774.
Le Masque de Fer [Alfred Wolff]. 'Échos de Paris'. *Le Figaro* (24 March 1875), 1.
Lostalot, Alfred de. 'Exposition des oeuvres de M. Claude Monet'. *Gazette des beaux-arts* (2nd Period) 27, no. 310 (April 1883), 342–348.
Major, Johann Daniel. 'Observation VIL: lumière bleue réfléchie par la neige'. In *Collection académique. Tome sixieme de la partie étrangère, et le premier de la physique expérimentale séparée*, 263, 1761.
Mazéas, Guillaume. 'Observations sur des couleurs engrendrées par le frottement des surfaces planes et transparentes'. In *Histoire de l'Académie Royale des Sciences et Belles-Lettres: années MDCCLII*, 248–261. Berlin: Haude et Spener, 1754.
Melvill, Thomas. *Essays and Observations, Physical and Literary*. Edinburgh: Hamilton and Balfour, 1756.
Millais, John Everett. *The Life and Letters of John Everett Millais*, edited by John Guille Millais. 2 vols. London: Methuen, 1899.
Milner, Isaac. 'Theory of Colours and Shadows'. In *Observations on the Theory and Practice of Landscape Gardening*, edited by Humphry Repton, 214–219. London: J. Taylor, 1803.
Monge, Gaspard. 'Mémoire sur quelques phénomènes de la vision'. *Annales de chimie* 3 (1789), 131–147.
Monge, Gaspard. 'Ueber einige Phänomene des Sehens'. *Journal der Physik* (edited by Albrecht Carl Gren) 2, no. 1 (1790), 142–154.
Monge, Gaspard. *Géométrie descriptive*. 4th edn. Paris: Courcier, 1820.

Müller, J. *Physiologie du système nerveux*. Translated by A.J.L. Jourdain. 2 vols. Paris: Baillère, 1840.

Newton, Isaac. *Opticks*. London: Benjamin Walford, 1704.

Nollet, Jean-Antoine. *Leçons de physique expérimentale*. 6 vols. Paris: Guerin and Delatour, 1743–64.

Ozanam, Jacques. *Récréations mathématiques et physiques*. 4 vols. Paris: Jombert, 1788.

Petrini, Pietro. 'Lettera del Sig. Ab. Pietro Petrini al Sig. Cioni sulle ombre colorate'. *Nuovo giornale dei letterati di Pisa* 2, no. 1 (January–February 1805), 45–48, and no. 2 (March–April 1805), 205–215.

Petrini, Pietro. 'Memoria sopra de' colori immaginarj nell'ombre'. *Nuovo giornale dei letterati di Pisa* 2, no. 3 (May–June 1805), 376–398.

Petrini, Pietro. 'Ricerche sulla produzione de colori immaginari nell'ombre'. *Memorie di matematica e fisica della società italiana delle scienze* 13, no. 2 (1807), 37–66.

Petrini, Pietro. *De' colori accidentale della luce*. Pistoia: Manfredi, 1815.

Priestley, Joseph. *The History and Present State of Discoveries Relating to Vision, Light, and Colours*. 2 vols. London: J. Johnson, 1772.

Prieur, Claude Antoine. 'Considérations sur les couleurs et sur plusieurs de leurs apparences'. *Annales de chimie* 54 (April 1805), 5–27.

Régnier, Jean-Désiré. *De la couleur et de la lumière chez les grands maîtres anciens*. Paris: Renouard, 1865.

Repton, Humphry. *Observations on the Theory and Practice of Landscape Gardening*. London: J. Taylor, 1803.

Repton, Humphry. *Fragments on the Theory and Practice of Landscape Gardening*. London: J. Taylor, 1816.

Reynolds, Joshua. *The Works of Sir Joshua Reynolds*, edited by Edmond Malone. 2nd edn. 2 vols. London: T. Cadell and W. Davies, 1798.

Richter, Henry. *Day-Light*. London: Ackerman, 1817.

Roget, John Lewis. *A History of the Old Water-Colour Society*. 2 vols. London and New York: Longmans, Green and Co., 1891.

Rood, Ogden N. *Modern Chromatics*. New York: Appleton, 1879.

Rossetti, William Michael, ed. *Pre-Raphaelite Diaries and Letters*. London: Hurst and Blackett, 1900.

Roux, Jacob. *Die Farben*. 3 vols. Heidelberg: C.F. Winter, 1824–29.

Runge, Philip Otto. *Farben-Kugel*. Hamburg: Friedrich Perthes, 1810.

Runge, Philip Otto. *Hinterlassene Schriften*, edited by Johann Daniel Runge, 168–170. 2 vols. Hamburg: Perthes, 1840.

Ruskin, John. *The Elements of Drawing*. 2nd edn. London: Smith, Elder, 1857.

Saint-Morien, M de. *La perspective aérienne*. Paris: Didot, 1788.

Sartain, John. *The Reminiscences of a Very Old Man 1808–1897*. New York: Appleton, 1899.

Saussure, Horace-Bénédict de. *Voyages dans les Alpes, précédés d'un essai sur l'histoire naturelle des environs de Genève*. 4 vols. Neuchâtel: Fauche-Borel, 1779–96.

Saussure, Horace-Bénédict de. 'Description d'un cyanomètre ou d'un appareil destiné à mesurer l'intensité de la couleur bleue du ciel'. *Journal de physique* 38 (March 1791), 199–208.

Scherffer, Karl. *Abhandlung von den zufälligen Farben*. Vienna: Johann Thomas Edlen von Trattnern, 1765.

Scherffer, Karl. 'Dissertation sur les couleurs accidentelles'. *Observations sur la physique, sur l'histoire naturelle, et sur les arts* 26, Part 1 (January and April 1785), 175–187 and 273–291.

Schreiber, Guido. *Die Farbenlehre*. Leipzig: Otto Spamer, 1868.

Signac, Paul. *De Delacroix au néo-impressionnisme*. Paris: Revue Blanche, 1899.

Silvestre, Theophile. *Eugène Delacroix: documents nouveaux*. Paris: Lévy, 1864.

Stephens, Frederick George. *William Holman Hunt and His Works*. London: James Nisbett, 1860.
Story, Alfred T. *James Holmes and John Varley*. London: Richard Bentley, 1894.
Sweetlove, John. 'The Natural Philosophy of Art'. *Art Journal* 4 (1852), 6–7 and 121–123.
Taine, Hippolyte. *De l'intelligence*. 2 vols. Paris: Hachette, 1870.
Thiebaut-Sisson. *Claude Monet: An Interview*. No Place of Publication: No Publisher, 1900.
Thompson, Benjamin, Count of Rumford. 'An Account of Some Experiments Upon Coloured Shadows'. *Philosophical Transactions of the Royal Society, London* 84 (1794), 107–118.
Thompson, Benjamin, Count of Rumford. 'An Account of Some Experiments Upon Coloured Shadows'. *A Journal of Natural Philosophy, Chemistry, and the Arts*, edited by William Nicholson, 1, no. 3 (June 1797), 101–106.
Thompson, Benjamin, Count of Rumford. 'An Account of Some Experiments Upon Coloured Shadows'. *The Annual Register for the Year 1794* 36 (1799), 342–348.
Thompson, Benjamin, Count of Rumford. 'Conjectures Respecting the Principles of the Harmony of Colours'. In *Philosophical Papers*, 333–340. London: Cadell, 1802.
Trimm, Timothée [Antoine Joseph Napoleon Lespès]. 'Dimanche 11 Décembre 1864: Les conférences d'Alexandre Dumas'. *Le petit journal* 681 (12 December 1864), 1–2.
Tupper, Jack [John]. 'Extracts From the Diary of an Artist'. *The Crayon* 2, no. 11 (12 September 1855), 159–160.
Tupper, Jack [John]. *Poems*, edited by William Michael Rossetti. London: Longmans, 1897.
Twining, Henry. *On the Philosophy of Painting*. London: Longman, Brown, Green, and Longmans, 1849.
Valenciennes, Pierre Henri de. *Elémens de perspective pratique*. Paris: The Author, Desene, and Duprat, An VIII, 1799 or 1800.
Vallée, Louis Léger. *Traite de géométrie descriptive*. Paris: Courcier, 1819.
Vallée, Louis Léger. *Traite de la science du dessin*. Paris: Courcier, 1821.
Venturi, Giambatista. 'Dei colori immaginari'. *Opuscoli scelto sulle scienze e sulle arte* 21 (1801), 274–288.
Venturi, Giambatista. *Indagine fisica sui colori*. 2nd edn. Modena: Società Tipographica, 1801.
Voigt, Johann Gottfried. 'Beobachtungen und Verſuche über farbigtes Licht, Farben und ihre Mischung'. *Neues Journal der Physik* 2, no. 2 (1796), 235–303.
Waern, Celia. *John La Farge*. London: Seeley and New York: Macmillan, 1896.
Walker, Adam. *A System of Familiar Philosophy*. London: The Author, and Kearsley and Edinburgh: Bell and Bradfute, 1799.
White, John. *Art's Treasury or Rarities and Curious Inventions*. 5th edn. London: G. Conyers, 1710.
Willson, Harry. *The Use of a Box of Colours*. London: Tilt and Bogue, 1842.
Wünsch, Christian. 'Bertrachtung des farbrigen Lichtes'. In *Kosmologische Unterhaltungen für die Jugend*, 272–292. Leipzig: Johann Cottl and Immanuel Breithopf, 1778.
Young, Thomas. 'An Account of Some Cases of the Production of Colours, Not Hitherto Described'. *Philosophical Transactions of the Royal Society of London* 92 (1802), 387–397.
Young, Thomas. *A Course of Lectures on Natural Philosophy and the Mechanical Arts*. 2 vols. London: Joseph Johnson, 1807.
Zola, Émile. 'Mon Salon IV: Les actualistes'. *L'Événement illustré* (24 May 1868).
Zola, Émile. *L'Oeuvre*. Paris: Charpentier, 1886.

Works Published After 1900

Albers, Josef. *The Interaction of Color*. 4th edn. New Haven, CT and London: Yale University Press, 2013.
Allen, Sue. 'Using Scientific Inquiry Activities in Exhibit Explanations'. *Science Education* 81, no. 6 (November 1997), 715–734.

Allert, Beate. 'Goethe, Runge, Friedrich: On Painting'. In *The Enlightened Eye: Goethe and Visual Culture*, edited by Evelyn K. More and Patricia Anne Simpson, 73–91. Amsterdam and New York: Rodopi, 2007.

Allert, Beate. '"Trübe" as the Source of New Color Formation in Goethe's Late Works *Entoptische Farben* (1817–20) and *Chromatik* (1822)'. *Goethe Yearbook* 19 (2012), 29–47.

Allert, Beate. 'J. W. Goethe and C. G. Carus: On the Representation of Nature in Science and Art'. *Goethe Yearbook* 23 (2016), 195–220.

Ames, Adelbert, C. A. Proctor, and Blanche Ames. 'Vision and the Technique of Art'. *Proceedings of the American Academy of Arts and Sciences* 58, no. 1 (February 1923), 3–47.

Amstutz, Nina. *Caspar David Friedrich*. New Haven, CT and London: Yale University Press, 2020.

Andersen, Jens. *Hans Christian Andersen*. Woodstock, NY and London: Overlook Duckworth, 2005.

Anon. 'De Saussure's Ascent of Mont Blanc'. *The Lotus Magazine* 5, no. 9 (June 1914), 577–586.

Aristotle. 'On Sense and Sensible Objects'. In *Parva Naturalia*. Translated by W.S. Hett, 215–283. Cambridge, MA, and London: Harvard University Press, 1957.

Aristotle. 'On Colours'. In *Minor Works*, translated by W.S. Hett, 5–45. Cambridge, MA and London: Harvard University Press, 2000.

Barris, Michael C. 'Barrow's Color Shadows'. *Optometry and Vision Science* 76, no. 7 (July 1999), 511–512.

Barrow, Isaac. *Isaac Barrow's Optical Lectures, 1667 (Lectiones XVIII)*, edited by A.G. Bennett and D.F. Edgar. Translated by H.C. Fay. London: The Worshipful Company of Spectacle Makers, 1987.

Baxandall, Michael. *Shadows and Enlightenment*. New Haven, CT and London: Yale University Press, 1995.

Bell, Janis. 'Aristotle as a Source for Leonardo's Theory of Colour Perspective after 1500'. *Journal of the Warburg and Courtauld Institutes* 56 (1993), 100–118.

Bendiner, Kenneth. *The Art of Ford Madox Brown*. University Park, PA: Pennsylvania State University Press, 1998.

Bennett, Mary. *William Holman Hunt*. Liverpool: Walker Art Gallery, 1969.

Bennett, Mary. *Artists of the Pre-Raphaelite Circle: The First Generation*. London: National Museums and Galleries on Merseyside, 1988.

Bennett, Mary. *Ford Madox Brown: A Catalogue Raisonné*. 2 vols. New Haven, CT and London: Yale University Press, 2010.

Benson, J.L. *Greek Color Theory and the Four Elements: A Cosmological Interpretation*. Amherst, MA: University of Massachusetts Amherst Libraries, 2000.

Bernard, Émile. 'Paul Cézanne'. *L'Occident* 6, no. 32 (July 1904), 17–30.

Berrie, B., F. Casadio, K. Dahm, Y. Strumfels, M. Tedeschi, and J. Walsh. 'A Vibrant Surface: Investigating Color, Texture and Transparency in Winslow Homer's Watercolors'. In *Science and Art: The Painted Surface*, edited by Antonio Sgamellotti, Brunetto Giovanni Brunetti, and Costanza Miliani, 404–428. Cambridge and London: The Royal Society of Chemistry, 2014.

Beucken, Jean de. *Un portrait de Cézanne*. Paris: Gallimard, 1955.

Blake, William. *The Complete Poetry and Prose of William Blake*, edited by David V. Erdman. Revised edn. New York: Doubleday, 1988.

Blakemore, C., and P. Sutton. 'Size Adaptation: A New Aftereffect'. *Science* 166, no. 3902 (1969), 245–247.

Bomford, David. 'The History of Colour in Art'. In *Colour, Art & Science*, edited by Trevor Lamb and Janine Bourriau, 7–30. Cambridge: Cambridge University Press, 1995.

Borbély, Ákos, Árpád Sámson, and János Schanda. 'The Concept of Correlated Colour Temperature Revisited'. *Color Research and Application* 26, no. 6 (December 2001), 450–457.

Borély, Jules. 'Cézanne à Aix'. *L'Art vivant* 2, no. 37 (1 July 1926), 491–493.

Börsch-Supan, Helmut. *Caspar David Friedrich*. London: Thames & Hudson, 1974.

Boskamp, Ulrike. 'Nachbilder, nicht komplementär Augenexperimente, Sehlüste und Modelle des Farbensehens im 18. Jahrhundert'. In *Nachbilder. Das Gedächtnis des Auges in Kunst und Wissenschaft*, edited by Werner Busch and Carolin Meister, 49–70. Zurich: Diaphanes, 2011.

Boyle, Nicholas. *Goethe: The Poet and the Age: Volume II: Revolution and Renunciation 1790–1803*. Oxford: Oxford University Press, 2000.

Braddick, Oliver. 'The Many Faces of Motion Perception'. In *The Artful Eye*, edited by Richard Gregory et al., 205–231. Oxford: Oxford University Press, 1995.

Brett, Charles. 'John Brett and Ruskin's "The Elements of Drawing"'. *The Burlington Magazine* 147, no. 1230 (September 2005), 613–616.

Bronkhurst, Judith. *William Holman Hunt: A Catalogue Raisonné*. 2 vols. New Haven, CT and London: Yale University Press, 2006.

Brou, Philippe, Thomas R. Sciascia, Lynette Linden, and Jerome Y. Lettvin. 'The Colors of Things'. *Scientific American* 255, no. 3 (September 1986), 84–91.

Brown, Ford Madox. *The Diary of Ford Madox Brown*, edited by Virginia Surtees. London: Yale University Press for the Paul Mellon Centre for Studies in British Art, 1981.

Bruce, Vicki, Patrick Green, and Mark Georgeson. *Visual Perception*. Hove: Psychology Press, 1996.

Bruno, Vincent J. *Form and Color in Greek Painting*. New York: W. W. Norton, 1977.

Burwick, Frederick. *The Damnation of Newton: Goethe's Color Theory and Romantic Perception*. Berlin and New York: de Gruyter, 2012.

Busch, Werner. 'Empirical Studies of Nature'. In *The Romantic Spirit in German Art 1790–1990*, edited by Keith Hartley, 278–282. London: Thames & Hudson, 1994.

Butlin, Martin, ed. *Samuel Palmer: The Sketchbook of 1824*. London: Thames & Hudson, 2005.

Callen, Anthea. *Techniques of the Impressionists*. London: Orbis, 1982.

Callen, Anthea. *The Art of Impressionism*. New Haven, CT and London: Yale University Press, 2000.

Carman, Taylor. *Merleau-Ponty*. London and New York: Routledge, 2008.

Carus, Carl Gustav. *Reise nach der insel Rügen*. Dresden: Wolfgang Jess, 1941.

Carus, Carl Gustav. *Nine Letters on Landscape Painting, Written in the Years 1815–1824*, edited by Oskar Bättschmann. Translated by David Britt. Los Angeles, CA: Getty Research Institute, 2002.

Cavina, Anna Ottina. *Paysages d'Italie. Les peintres du plein air (1780–1830)*. Paris: Réunion des Musées Nationaux, 2001.

Charnay, Yves, and Eckhart Sieffert. 'Les ombres colorées'. *Pour la science* 289 (November 2001), 104–105.

Charteris, Evan. *John Sargent*. New York: Charles Scribner, 1927.

Christensen, Dan Charly. *Hans Christian Ørsted*. Oxford: Oxford University Press, 2013.

Churma, Michael E. 'Blue Shadows: Physical, Physiological, and Psychological Causes'. *Applied Optics* 33, no. 4 (July 1994), 4719–4722.

Clémenceau, Georges. *Claude Monet: les nymphéas*. Paris: Plon, 1928.

Cohen, Jonathan. *The Red and the Real: An Essay on Colour Ontology*. Oxford: Oxford University Press, 2009.

Cohen, Jozef. *Visual Color and Color Mixture*. Urbana and Chicago, IL: University of Illinois Press, 2001.

Conisbee, Philip. 'Eckersberg, an Original in His Time'. In Philip Conisbee et al., *Christoffer Wilhelm Eckersberg 1783–1853*, 26–53. Washington, DC: National Gallery, 2003.

Conisbee, Philip, Kasper Monrad, and Lene Bogh Ronberg. *Christoffer Wilhelm Eckersberg 1783–1853*. Washington, DC: National Gallery, 2003.

Coombs, James H., Anne M. Scott, George P. Landow, and Arnold A. Sanders, eds. *A Pre-Raphaelite Friendship: The Correspondence of William Holman Hunt and John Lucas Tupper*. Ann Arbor, MI: UMI Research Press, 1986.

Cooper, Helen A. *Winslow Homer Watercolours*. Washington, DC: National Gallery of Art, 1986.

Cortissoz, Royal. *John La Farge: A Memoir and a Study*. Boston, MA and New York: Houghton Mifflin, 1911.
Crone, Robert A. *A History of Color*. Amsterdam: Kluwer, 1999.
da Vinci, Leonardo. *The Literary Works of Leonardo da Vinci*, edited by Jean-Paul Richter. 2nd edn. London: Oxford University Press, 1939.
da Vinci, Leonardo. *Treatise on Painting (Codex Urbinas Latinus 1270)*. Translated and Annotated by Philip McMahon. 2 vols. Princeton, NJ: Princeton University Press, 1956.
Davis, Raymond. 'A Correlated Color Temperature for Illuminants'. *Bureau of Standards Journal of Research* 7 (January 1931), 659–681.
Denis, Maurice. *Théories 1890–1910*. Paris: L'Occident, 1912.
De Valois, Russell L., and Karen K. De Valois. 'Neural Coding of Colour'. In *The Science of Color: Vol. 2 of Readings on Color*, edited by Alex Byrne and David R. Hilbert, 94–140. Cambridge, MA: MIT Press, 1997.
De Vries, Lyckle. 'Gerard de Lairesse: The Critical Vocabulary of an Art Theorist'. *Oud Holland* 117, nos 1–2 (2004), 79–98.
Dewhurst, Wynford. *Impressionist Painting: Its Genesis and Development*. London: George Newness, 1904.
Dewhurst, Wynford. 'What Is Impressionism?'. *Contemporary Review* 99 (March 1911), 286–302.
Dorra, Henri. 'Valenciennes's Theories: From Newton, Buffon and Diderot to Corot, Chevreul, Delacroix, Monet and Seurat'. *Gazette des beaux-arts* (6th Period) 124, no. 1510 (November 1994), 185–194.
Duck, Michael J. 'Newton and Goethe on Colour'. *Annals of Science* 45, no. 5 (1988), 507–519.
Dulon, Guy, and Christophe Duvivier. *Louise Hayet*. Pontoise: Musée de Pontoise, 1991.
Ehrenzweig, Anton. *The Psycho-Analysis of Artistic Vision and Hearing*. London: Routledge and Kegan Paul, 1953.
Escholier, Raymond. *Delacroix: peintre, graveur, écrivain*. 3 vols. Paris: Floury, 1926.
Evans, Gareth. *The Varieties of Reference*, edited by John McDowell. Oxford: Clarendon Press and New York: Oxford University Press, 1982.
Evans, Ralph M. *An Introduction to Color*. New York: John Wiley and London: Chapman & Hall, 1948.
Farago, Claire, Janis Bell, and Carlo Vecce. *The Fabrication of Leonardo da Vinci's Trattato della pittura*. 2 vols. Boston, MA and Leiden: Brill, 2018.
Fine, Ione. 'Recovery of Vision Following Blindness'. In *Encyclopedia of Perception*, edited by E. Bruce Goldstein, 864–866. Thousand Oaks, CA: Sage, 2010.
Finley, Gerald E. 'Turner: An Early Experiment With Colour Theory'. *Journal of the Warburg and Courtauld Institutes* 30 (1967), 357–366.
Fiorani, Francesca. 'The Colors of Leonardo's Shadows'. *Leonardo* 41, no. 3 (2008), 271–278.
Fiorani, Francesca. 'The Shadows of Leonardo's Annunciation and Their Lost Legacy'. In *Imitation, Representation and Printing in the Italian Renaissance*, edited by Roy Eriksen and Magne Malmanger, 119–156. Pisa and Rome: Fabrizio Serra, 2009.
Fisher, Philip. *Wonder, the Rainbow, and the Aesthetics of Rare Experiences*. Cambridge, MA: Harvard University Press, 2003.
Fodor, Jerry A., and Zenon W. Pylyshyn. 'How Direct Is Visual Perception? Some Reflections on Gibson's "Ecological Approach"'. *Cognition* 9 (1981), 139–196.
Foster, David. 'Colour Constancy'. *Vision Research* 51, no. 7 (2011), 674–700.
Foster, David, and Sérgio Nascimento. 'Relational Colour Constancy from Invariant Cone-Excitation Ratios'. *Proceedings of the Royal Society B: Biological Sciences* 257 (1994), 115–121.
Foster, Kathleen A. 'The Still-Life Painting of John La Farge'. *The American Art Journal* 11, no. 3 (July 1979), 4–37.

Fraser, David. 'Joseph Wright of Derby and the Lunar Society'. In *Wright of Derby*, edited by Judy Egerton, 15–23. London: Tate Gallery, 1990.
Gage, John. *A Decade of English Naturalism 1810–1820*. Hunstanton: Witley Press, 1969.
Gage, John. *George Field and His Circle: From Romanticism to the Pre-Raphaelite Brotherhood*. Cambridge: Fitzwilliam Museum, 1989.
Gage, John. *Colour and Culture*. London: Thames & Hudson, 1993.
Gage, John. 'A Romantic Colourman: George Field and British Art'. *The Volume of the Walpole Society* 63 (2001), 1–73.
Gage, John. *Colour in Art*. London: Thames & Hudson, 2006.
Gallo, Luigi. *Pierre-Henri de Valenciennes 1750–1819: l'artiste et le théoricien*. Rome: "L'Erma" di Bretschneider, 2017.
Garver, Newton. *This Complicated Form of Life*. Peru, IL: Open Court, 1994.
Gascoigne, Bamber. *Milestones in Colour Printing 1457–1859*. Cambridge: Cambridge University Press, 1997.
Gasquet, Joachim. *Cézanne*. Paris: Bernheim-Jeune, 1921.
Gedo, Mary Matthews. *Monet and His Muse*. Chicago, IL: The University of Chicago Press, 2010.
Geffroy, Gustave. *Claude Monet, sa vie, son temps, son oeuvre*. Paris: G. Crès, 1924.
Gelb, Adhémar. 'Die "Farbenkonstanz" der Sehrdinge'. *Handbuch der Normalen und Pathologischen Physiologie* 12, no. 1 (1929), 594–678.
Glanville, Helen. 'Contemporary Colour Theory'. In *Pre-Raphaelite Painting Techniques, 1848–56*, edited by Joyce H. Townsend, Jaqueline Ridge, and Stephen Hackney, 29–37. London: Tate, 2004.
Good, Justin. *Wittgenstein and the Theory of Perception*. London: Continuum, 2006.
Goodrich, Lloyd. *Winslow Homer*. New York: Macmillan, 1944.
Gottschalk, H.B. 'The De Coloribus and Its Author'. *Hermes* 92, no. 1 (1964), 59–85.
Grave, Johannes. *Caspar David Friedrich*. New York: Prestel, 2012.
Gray, Ronald. *Goethe the Alchemist*. Cambridge: Cambridge University Press, 1952.
Gruetzner, Anna. *A Fragile Modernism: Whistler and His Impressionist Followers*. New Haven, CT and London: Yale University Press, 2007.
Hansen, Peter H. *The Summits of Modern Man: Mountaineering After the Enlightenment*. Cambridge, MA: Harvard University Press, 2013.
Hansen, Thorsten, Lars Pracejus, and Karl Gegenfurtner. 'Color Perception in the Intermediate Periphery of the Visual Field'. *Journal of Vision* 9, no. 26 (April 2009), 1–12.
Hanson, Norwood Russel. *The Concept of the Positron*. Cambridge: Cambridge University Press, 1963.
Hanson, Norwood Russel. *Patterns of Discovery*. 3rd edn. Cambridge: Cambridge University Press, 1965.
Hardin, C.L. *Color for Philosophers*. Indianapolis, IN and Cambridge: Hackett, 1988.
Harmon, Leon D., and Bela Julez. 'Masking in Visual Recognition'. *Science* 180 (15 June 1973), 1194–1196.
Harmon, Leon D. 'The Recognition of Faces'. *Scientific American* 229, no. 5 (November 1973), 70–83.
Hedin, Gry. 'Image Construction'. In *Christoffer Wilhelm Eckersberg*, edited by Kapser Monrad et al., 151–172. Munich, London, and New York: Prestel, 2015.
Heinstein, Patrick. 'Roux, Jacob'. In *Neue Deutsche Biographie*. Vol. 22, 148–149. Berlin: Duncker & Humblot, 2005. www.deutsche-biographie.de/pnd116665505.html#ndbcontent
Helmholtz, Hermann von. *Physiological Optics*. 2 vols. 2nd edn. New York: Dover, 1962.
Herbert, Robert L., Françoise Cachin, Anne Distel, and Gary Tinterow. *Seurat*. Paris: Réunion des Musees Nationaux, 1991.
Hering, Ewald. *Outlines of a Theory of the Light Sense*. Translated by Leo M. Hurvich and Dorothea Jameson. Cambridge, MA: Harvard University Press, 1964.

Hickox, Michael. 'John Brett and Ruskin'. *The Burlington Magazine* 138, no. 1121 (August 1996), 521–525.

Hilton, Timothy. *The Pre-Raphaelites*. London: Thames & Hudson, 1970.

Hoeppe, Götz. *Why Is the Sky Blue*. Translated with John Stewart. Princeton, NJ and Oxford: Princeton University Press, 2007.

Hoermann, Kirstin. 'A Hand Formed to Use the Brush'. In *Winslow Homer: Paintings of the Civil War*, edited by Marc Simpson, Nicolai Cikovsky Jr., Lucretia Hoover Giese, Kirsten Hoermann, Sally Mills, and Christopher Kent Wilson, 103–119. San Francisco, CA: The Fine Arts Museums of San Francisco and San Francisco: Bedford Arts, 1988.

Holmes, John. 'Pre-Raphaelitism, Science and the Arts in "the Germ"'. *Victorian Literature and Culture* 43, no. 4 (2015), 689–703.

Homer, William Innes. *Seurat and the Science of Painting*. Cambridge, MA: Massachusetts Institute of Technology, 1978.

House, John. *Monet: Nature Into Art*. New Haven, CT and London: Yale University Press, 1986.

Houston, George W. 'Tiberius on Capri'. *Greece & Rome* 32, no. 2 (October 1985), 179–196.

Hueffer, Ford Madox. *Ancient Lights and Certain New Reflections, Being the Memoirs of a Young Man*. London: Chapman & Hall, 1911.

Hunt, William Holman. *Pre-Raphaelitism and the Pre-Raphaelite Brotherhood*. 2 vols. London: Macmillan, 1905.

Hurlbert, Anya. 'Computational Models of Colour Constancy'. In *Why Things Look as They Do*, edited by Vincent Walsh and Janusz Kulikowski, 283–322. Cambridge: Cambridge University Press, 1998.

Hurlbert, Anya. '"Colour Vision": Primary Visual Cortex Shows Its Influence'. *Current Biology* 13, no. 7 (April 2003), 270–272.

Hurlbert, Anya, and Kit Wolf. 'Color Contrast: A Contributory Mechanism to Color Constancy'. *Progress in Brain Research* 144 (February 2004), 147–160.

Hurvich, Leo M. 'Chromatic and Achromatic Response Functions'. In *The Science of Color: Vol. 2 of Readings on Color*, edited by Alex Byrne and David R. Hilbert, 68–91. Cambridge, MA: MIT Press, 1997.

Hurvich, Leo M., and Jameson, Dorothea. 'An Opponent Process Theory of Color Vision'. *Psychological Review* 64, no. 6 (November 1957), 384–404.

Ivins, William M. *Prints and Visual Communication*. Cambridge, MA: Harvard University Press, 1953.

Jacob, Pierre, and Marc Jeannerod. *Ways of Seeing: The Scope and Limits of Visual Cognition*. Oxford: Oxford University Press, 2003.

John, Barbara. *Caspar David Freidrich: Kriedefelsen au Rügen*. Leipzig: Seeman, 2005.

Johnson, Galen, ed. *The Merleau-Ponty Aesthetics Reader*. Evanston, IL: Northwestern University Press, 1993.

Johnson, Lee. *Delacroix*. London: Weidenfeld & Nicholson, 1963.

Johnson, Lee. *The Paintings of Eugène Delacroix: A Critical Catalogue*. 6 vols. Oxford: Clarendon Press, 1981–89.

Johnson, Lee. 'Some Sketches Made by Delacroix at Dieppe'. *The Burlington Magazine* 143, no. 1177 (April 2001), 224–227.

Joost, Ulrich. '"Da ich . . . Den bunten Schattennachlaufe, wie ehmals als Knabe den Schmetterlingen" – Lichtenberg, Goethe, and the Coloured Shadows'. *Color Research and Application* (supplement volume) 26 (January 2001), 28–31.

Joost, Ulrich, Barry Lee, and Qasim Zaidi. 'Lichtenberg's Letter to Goethe on "Färbige Schatten"'. *Color Research and Application* 27, no. 4 (June 2002), 300–303.

Kallmann, Julian, Jürgen Schramme, and Christa Neumeyer. 'The Phenomenon of Coloured Shadows'. In *Perception Beyond Gestalt*, edited by Adam Geremeck, Mark W. Greenlee, and Svein Magnussen, 185–197. London: Psychology Press, 2014.

Kant, Immanuel. *Critique of Judgment*. Translated by James Creed Meredith. Oxford: Oxford University Press, 2007 [1790].

Katz, David. *The World of Colour*. Translated by R.B. McLeod. London: Kegan Paul, Trench, Trubner & Co., 1935.

Katz, Melissa R. 'William Holman Hunt and the "Pre-Raphaelite Technique"'. In *Historical Painting Techniques, Materials, and Studio Practice*, edited by Arie Walert, Erna Hermens, and Marja Peek, 158–165. Los Angeles, CA: The J. Paul Getty Trust, 1995.

Kelly, Sean Dorrance. 'Seeing Things in Merleau-Ponty'. In *The Cambridge Companion to Merleau-Ponty*, edited by Taylor Carman and Mark B.N. Hansen, 74–110. Cambridge: Cambridge University Press, 2005.

Kemp, Martin. *The Science of Art: Optical Themes in Western Art from Brunelleschi to Seurat*. New Haven, CT and London: Yale University Press, 1990.

Kemp, Martin. 'The "Hammer Lecture" (1992)'. In the Beholder's Eye: Leonardo and the "Errors of Sight" in Theory and Practice'. In *Leonardo's Science and Technology: Essential Readings for the Non-Scientist*, edited by Claire Farago, 155–163. New York: Garland, 1999.

Kessler, Count Harry. *Journey to the Abyss: The Diaries of Count Harry Kessler, 1880–1918*, edited by Laird Easton. New York: Knopf, 2011.

Kessler, Count Harry. *Journal, regards sur l'art et les artistes contemporains, 1889–1937*, edited by Ursel Berger, Julia Drost, Alexandre Kostka, Antoinette Le Normand-Romain, Dominique Lobstein, and Philippe Thiébaut. Translated by Jean Torrent. Paris: Éditions de la maison des sciences de l'homme, 2017. https://books.openedition.org/editionsmsh/10918

Kittelmann, Udo. 'Fliegt Geister, fleigt, erregt die Luft . . . Wer bitte? August Kopisch?'. In *August Kopisch: Maler, Dichter, Entdecker, Erfinde*, edited by Udo Kittelmann et al., 13–17. Dresden: Sandstein Verlag, 2016.

Kittelmann, Udo, Franziska Lietzmann, Sabina Magnani von Petersdorff, Andreas Platthaus, and Dieter Richter. *August Kopisch: Maler, Dichter, Entdecker, Erfinder*. Dresden: Sandstein Verlag, 2016.

Kobbé, Gustave. 'John La Farge and Winslow Homer'. *New York Herald* (4 December 1910), Magazine, 11.

Koenderink, Jan, Andrea van Doorn, Christoph Witzel, and Karl Gegenfurtner. 'Hues of Color Afterimages'. *I-Perception* 11, no. 1 (January 2020), 1–18.

Kopisch, August. *The Discovery of the Blue Grotto*. Translated by Veronica Galbiati. Naples: Intra Moenia, 2016.

Kuehni, Rolf G. 'Memoir Concerning Certain Phenomena of Vision by M. Monge'. *Color Research and Application* 22, no. 3 (June 1997), 199–203.

Kuhn, Thomas. *The Structure of Scientific Revolutions*. 4th edn. Chicago, IL and London: The University of Chicago Press, 2012.

La Farge, John. 'On Painting'. *New England Magazine* 38 (April 1908), 229–240.

Land, Edwin H. 'Experiments in Color Vision'. *Scientific American* 200, no. 5 (May 1959), 84–99.

Land, Edwin H. 'The Retinex Theory of Color Vision'. *Scientific American* 237, no. 6 (1977), 108–128.

Land, Edwin H. 'Recent Advances in Retinex Theory and Some Implications for Cortical Computations'. *Proceedings of the National Academy of Sciences of the United States of America* 80, no. 16 (August 1983), 5163–5169.

Land, Edwin H., and John J. McCann. 'Lightness and Retinex Theory'. *Journal of the Optical Society of America* 61, no. 1 (January 1971), 1–11.

Lanthony, Phillippe. *Des yeux pour peindre*. Paris: Réunion des Musées Nationaux, 2006.

Lavezzi, Elizabeth. 'Peinture et savoirs scientifiques'. *Dix-huitième siècle* 31 (1999), 233–247.

Lichtenberg, Georg Christophe. *Philosophical Writings*. Edited and translated by Steven Tester. New York: State University of New York Press, 2012.

Lister, Raymond. *Catalogue Raisonné of the Works of Samuel Palmer*. Cambridge: Cambridge University Press, 1988.
Livingstone, Margaret. *Vision and Art: The Biology of Seeing*. New York: Harry N. Abrams, 2002.
Lowengard, Sarah. *The Creation of Color in Eighteenth-Century Europe*. New York: Columbia University Press, 2006. www.gutenberg-e.org/lowengard/
Lui, Francesca. 'La Sicilia nell'occhio del pittore: il viaggio del 1779 di Pierre-Henri de Valenciennes'. In *La Sicilia e il Grand Tour: la riscoperta di Akragas 1700–1800*, edited by Alessandro Carlino, 195–208. Rome: Gangemi, 2009.
MacColl, Dugald Sutherland. *Nineteenth Century Art*. Glasgow: James Maclehose & Sons, 1902.
MacColl, Dugald Sutherland. 'Impressionism'. In *Encyclopaedia Brittanica*. 10th edn. Vol. 30, 413–415. London: A. & C. Black, 1902–03.
MacMillan, John Duncan. 'Holman Hunt's Hireling Shepherd'. *The Art Bulletin* 54, no. 2 (June 1972), 187–197.
Margócsy, Dániel. *Commercial Visions*. Chicago, IL: The University of Chicago Press, 2014.
Mather, George. *The Psychology of Visual Art*. Cambridge: Cambridge University Press, 2013.
Matthaei, Rupprecht, ed. *Goethe's Colour Theory*. London: Studio Vista, 1971.
Matthen, Mohan. 'How Things Look (and What Things Look That Way)'. In *Perceiving the World*, edited by Bence Nanay, 239–253. Oxford: Oxford University Press, 2010.
McCollough, C. 'Color Adaptation of Edge-Detectors in the Human Visual System'. *Science* 149, no. 3688 (September 1965), 1115–1116.
McDowell, John. *Mind and World*. Cambridge, MA and London: Harvard University Press, 1994.
McGinn, Marie. *Wittgenstein and the Philosophical Investigations*. London: Routledge, 1977.
Merleau-Ponty, Maurice. *The Visible and the Invisible*. Evanston, IL: Northwestern University Press, 1968.
Merleau-Ponty, Maurice. *Phenomenology of Perception*. London and New York: Routledge, 1996 [1962].
Milner, A. David, and Melvyn A. Goodale. *The Visual Brain in Action*. Oxford: Oxford University Press, 1995.
Minnaert, Marcel. *Light and Colour in the Open Air*. Translated by H.M. Kremer-Priest and K.E. Brian Jay. New York: Dover, 1954.
Mirbeau, Octave. *Dans le ciel*. Paris: Editions du Boucher, 2003 [1892–93].
Mollon, John D. 'The Origins of Modern Colour Science'. In *The Science of Color*, edited by Steven Shevell, 2–39. Amsterdam: Elsevier, 2003.
Mollon, John. 'Monge: The Verriest Lecture, Lyon, July 2005'. *Visual Neuroscience* 23, nos 3–4 (July 2006), 297–309.
Monrad, Kasper. 'Eckersberg and Open-Air Painting'. In *Christoffer Wilhelm Eckersberg 1783–1853*, edited by Philip Conisbee et al., 16–25. Washington, DC: National Gallery, 2003.
Monrad, Kasper. 'Eckersberg on the European Stage'. In *Christoffer Wilhelm Eckersberg*, edited by Kapser Monrad et al., 9–54. Munich, London, and New York: Prestel, 2015.
Monrad, Kapser, Henrik Holm, Anna Schram Vejlby, Neela Struck, Jesper Svenningsen, Gry Hedin, and David Jackson. *Christoffer Wilhelm Eckersberg*. Munich, London, and New York: Prestel, 2015.
Moran, J. W. 'Blue Shadows in Nature and Art'. *The International Studio: An Illustrated Magazine of Fine and Applied Art* 40, no. 159 (May 1910), 74–77 and 41, no. 163 (September 1910), 76–78.
Mortier, Edouard, Duc de Trévise. 'Le pèlerinage de Giverny'. *La revue des l'art ancien et moderne* 51 (January–May 1927), 121–134.
Muenzer, Clark S. 'Fugitive Images and Visual Memory in Goethe's Discourse on Colour'. In *The Enlightened Eye: Goethe and Visual Culture*, edited by Evelyn K. More and Patricia Anne Simpson, 219–237. Amsterdam and New York: Rodopi, 2007.

Musson, Albert Edward and Eric Robinson. *Science and Technology in the Industrial Revolution*. Manchester: Manchester University Press, 1969.

Nicolson, Benedict. *Joseph Wright of Derby: Painter of Light*. 2 vols. London: Routledge & Kegan Paul; New York: Pantheon, 1968.

O'Keefe, Paul. *A Genius for Failure: The Life of Benjamin Robert Haydon*. London: Bodley Head, 2009.

Olivieri, G., G. Torosantucci, and M. Vicentini. 'Coloured Shadows'. *International Journal of Science Education* 10, no. 5 (1988), 561–569.

O'Regan, Kevin and Alva Noë. 'A Sensorimotor Account of Vision and Visual Consciousness'. *Behavioral and Brain Sciences* 24, no. 5 (November 2001), 939–973.

Ott, Gerhard. *Goethe, Eckermann, und die farbigen Schatten*. Dornach: Goethe-Farbstudio, 1977.

Palmer, Samuel. *The Letters of Samuel Palmer*, edited by Raymond Lister. 2 vols. Oxford: Clarendon Press, 1974.

Parris, Leslie, ed. *The Pre-Raphaelites*. London: Tate, 1984.

Payne, Christiana. *John Brett, Pre-Raphaelite Landscape Painter*. New Haven, CT and London: Yale University Press, 2012.

Peacocke, Christopher. *A Study of Concepts*. Cambridge, MA, and London: MIT Press, 1992.

Pearce, Bradley, Stuart Crichton, Michael Mackiewitz, Graham D. Finlayson, and Anya Hurlbert. 'Chromatic Illumination Discrimination Ability Reveals That Human Colour Constancy Is Optimised for Blue Daylight Illuminations'. *PLOS One* 9, no. 2 (February 2014), e87989.

Pedretti, Carlo. *Leonardo Da Vinci on Painting: A Lost Book (Libro A) Reassembled From the Codex Vaticanus Urbinas 1270 and From the Codex Leicester*. Berkeley and Los Angeles, CA: University of California Press, 1964.

Perry, Lilla Cabot. 'Reminiscences of Claude Monet from 1889–1909'. *The American Magazine of Art* 18, no. 3 (1927), 119–126.

Petersdorff, Sanina Magnani von. 'August Kopisch und die italienische Sprache un Literatur'. In *August Kopisch: Maler, Dichter, Entdecker, Erfinde*, edited by Udo Kittelmann et al., 47–53. Dresden: Sandstein Verlag, 2016.

Pissarro, Camille. *Correspondance de Camille Pissarro*, edited by Janine Bailly-Herzberg. 5 vols. Paris: Presses Universitaires de France, and Valhermeil, 1980–91.

Pomarède, Vincent. '1821–34: The Making of An Artist'. In *Corot*, edited by Michael Pantazzi, Vincent Pomarède, and Gary Tinterow. New York: Metropolitan Museum of Art, 1996.

Prettejohn, Elizabeth. *The Art of the Pre-Raphaelites*. London: Tate, 2000.

Priest, Irwin. 'Blue Sky and White Snow'. *Journal of the Optical Society of America and Review of Scientific Instruments* 13 (1926), 308.

Raffman, Diana. 'On the Persistence of Phenomenology'. In *Conscious Experience*, edited by Thomas Metzinger, 292–308. Paderborn: Schöningh and Thorverton: Imprint Academic, 1995.

Ratcliffe, Robert. *Cézanne's Working Methods and Their Theoretical Background*. PhD diss., University of London, 1960.

Renoir, Jean. *Renoir*. Paris: Hachette, 1962.

Reutersvärd, Oscar. 'The "Violettomania" of the Impressionists'. *The Journal of Aesthetics and Art Criticism* 9, no. 2 (December 1950), 106–110.

Rewald, John. *The History of Impressionism*. 4th edn. London: Secker & Warburg, 1973.

Rewald, John. *Paul Cézanne*. London: Thames & Hudson, 1986.

Richter, Dieter. 'Dichter, Kunstler, Lenenskünstler: August Kopischs romantische Existenz'. In *August Kopisch*, edited by Udo Kittelmann et al., 31–45. Dresden: Sandstein Verlag, 2016.

Rivière, R.P., and Jacques Schnerb. 'L'Atelier de Cézanne'. *La grande revue* 46 (1907), 811–817; Reprinted in *Conversations avec Cézanne*, edited by Michal Doran, 85–91. Paris: Macula, 1978.

Robinson, Eric. 'The Lunar Society'. *Transactions of the Newcomen Society* 35, no. 1 (1962), 153–178.

Roque, Georges. 'Les couleurs complémentaires; un nouveau paradigme'. *Revue d'histoire des sciences* 47, nos 3–4 (July–December 1994), 405–434.
Roque, Georges. 'Chevreul and Impressionism: A Reappraisal'. *The Art Bulletin* 78, no. 1 (March 1996), 26–39.
Roque, Georges. *Art et science de la couleur: Chevruel et les peintres, de Delacroix à l'abstraction.* 2nd edn. Paris: Gallimard, 2009.
Roux, Marius. *The Substance and the Shadow*, edited by Paul Smith. Translated by Dick Collins and Fiona Cox. University Park, PA: Pennsylvania State University Press, 2007.
Rovi, Alberto. 'Il "Saggio di ricerche intorno all'armonia cromatica naturale ed artificiale" di Giuseppe Bossi'. *Arte lombarda*, New Series, nos 55–57 (1980), 373–385.
Ruston, Sharon, 'Shelley's Links to the Midlands Enlightenment: James Lind and Adam Walker'. *British Journal for Eighteenth-Century Studies* 30, no. 2 (2007), 227–241.
Schaeffer, Iris, Caroline von Saint-George, and Katja Lewerentz. *Painting Light: The Hidden Techniques of The Impressionists*. Milan: Skira, 2008.
Schofield, Robert E. *The Lunar Society of Birmingham: A Social History of Provincial Science and Industry in Eighteenth-Century England*. Oxford: Clarendon Press, 1963.
Schofield, Robert E. 'The Lunar Society of Birmingham: A Bicentenary Appraisal'. *Notes and Records of the Royal Society of London* 21, no. 2 (December 1966), 144–161.
Schöller, Julia. *Schatten: Zur Malerei im 19. Jahrhundert*. Berlin and Munich: Deutscher Kunstverlag, 2017.
Schyns, Philippe G., and Oliva Aude. 'Dr. Angry and Mr. Smile: When Categorization Flexibly Modifies the Perception of Faces in Rapid Visual Presentations'. *Cognition* 69, no. 3 (January 1999), 243–265.
Sepper, Dennis L. *Goethe Contra Newton: Polemics and the Project for a New Science of Colour*. Cambridge: Cambridge University Press, 1988.
Sérusier, Paul. *ABC de la peinture*. Paris: Floury, 1942.
Shank, J.B. *The Newton Wars and the Beginning of the French Enlightenment*. Chicago, IL and London: The University of Chicago Press, 2008.
Shiff, Richard. 'Cézanne's Physicality: The Politics of Touch'. In *The Language of Art History*, edited by Salim Kemal and Ivan Gaskell, 129–180. Cambridge: Cambridge University Press, 1991.
Siegel, Linda. *Caspar David Friedrich and the Age of German Romanticism*. Boston, MA: Branden Books, 1978.
Sik-Lanyi, Cecilia, Veronika Szücs, and Robert Hirschler. 'Coloured Shadows: Why They Can Be Photographed'. *Color Research and Application* 44, no. 6 (December 2019), 859–874.
Sjåstad, Øystein. *A Theory of the Tache in Nineteenth-Century Painting*. Farnham: Ashgate, 2014.
Smith, Alison. 'The Enfranchised Eye'. In *Pre-Raphaelite Vision: Truth to Nature*, edited by Allen Staley, Christopher Newall, Alison Smith, Ian Warrell, and Tim Batchelor, 11–21. London: Tate, 2004.
Smith, Alison. 'Medium and Method in Pre-Raphaelite Painting'. In *Pre-Raphaelites: Victorian Avant-Garde*, edited by Tim Barringer, Jason Rosenfeld, Alison Smith, Elizabeth Prettejohn, and Dianne Waggoner, 18–23. London: Tate, 2012.
Smith, Paul. '"Parbleu": Pissarro and the Political Colour of an Original Vision'. *Art History* 15, no. 2 (June 1992), 223–247.
Smith, Paul. 'Cézanne's "Primitive" Perspective, or the "View from Everywhere"'. *The Art Bulletin* 95, no. 1 (March 1993), 102–119.
Smith, Paul. *Seurat and the Avant-Garde*. New Haven, CT and London: Yale University Press, 1997.
Smith, Paul. 'Cézanne's Colour Lab: Not-So-Still Life'. In *The World is an Apple: The Still Lifes of Paul Cézanne*, edited by Benedict Leca, 92–144. Hamilton: Art Gallery of Hamilton and London: Giles, 2014.

Smith, Paul. '"The Most Beautiful Blue": Painting, Science, and the Perception of Coloured Shadows'. *British Journal of Aesthetics* 58, no. 4 (October 2018), 385–400.

Sorensen, Roy. *Seeing Dark Things: The Philosophy of Shadows*. Oxford: Oxford University Press, 2008.

Sparti, Donatella Livia. 'Cassiano Dal Pozzo, Poussin and the Making and Publication of Leonardo's *Trattato*. *Journal of the Warburg and Courtauld Institute* 66 (2003), 143–188.

Spillmann, Lothar, and John S. Werner. 'Long-Range Interactions in Visual Perception'. *Trends Neuroscience* 19, no. 10 (October 1996), 428–434.

Staley, Allen. *The Pre-Raphaelite Landscape*. Oxford: Clarendon Press, 1973.

Staley, Allen, Christopher Newall, Alison Smith, Ian Warrell, and Tim Batchelor. *Pre-Raphaelite Vision: Truth to Nature*. London: Tate, 2004.

Strick, Jeremy. 'Nature Studied and Nature Remembered: The Oil Sketch in the Theory of Pierre-Henri de Valenciennes'. In *In the Light of Italy: Corot and Early Open-Air Painting*, edited by Philip Conisbee, Sarah Faunce, Peter Galassi, Vincent Pomarède, and Jeremy Strick, 79–87. Washington, DC: National Gallery of Art, 1996.

Stuckey, Charles. 'Monet's Art and the Act of Vision'. In *Aspects of Monet: A Symposium on the Artist's Life and Times*, edited by John Rewald and Frances Weitzenhoffer, 106–121. New York: Harry N. Abrams, 1984.

Tantillo, Astrida Orle. *The Will to Create: Goethe's Philosophy of Nature*. Pittsburgh, PA: University of Pittsburgh Press, 2002.

Tatham, David. 'Winslow Homer's Library'. *The American Art Journal* 9, no. 1 (May 1977), 92–98.

Taylor, Paul. 'The Concept of *Houding* in Dutch Art Theory'. *Journal of the Warburg and Courtauld Institutes* 55 (1992), 210–232.

Tedeschi, Raffaella. *L'impressionismo di Octave Mirbeau: un'estetica del fluido*. PhD diss., Università degli studi di Catania, 2012, 174. http://archivia.unict.it/bitstream/10761/1235/1/TDSRFL77C41C351N-These.pdf

Thirlwell, Angela. 'From Paint to Print: Grandfather's Legacy'. In *Ford Madox Brown and Visual Culture*, edited by Laura Colombino, 29–38. Amsterdam and New York: Rodopi, 2009.

Thorold, Anne. *Artists, Writers, Politics: Camille Pissarro and His Friends*. Oxford: Ashmolean Museum, 1980.

Townsend, Joyce H., Jaqueline Ridge, and Stephen Hackney, eds. *Pre-Raphaelite Painting Techniques, 1848–56*. London: Tate, 2004.

Treuherz, Julian, Angela Thirlwell, and Kenneth Bendiner. *Ford Madox Brown: Pre-Raphaelite Pioneer*. London: Philip Wilson, 2011.

Uglow, Jenny. *The Lunar Men: The Friends Who Made the Future 1730–1810*. London: Faber & Faber, 2002.

Valberg, Arne. *Light Vision Color*. Chichester: John Wiley & Sons, 2005.

Vanderpoel, Emily Noyes. *Color Problems*. 3rd edn. New York, London, and Bombay: Longmans, Green & Co., 1903.

Van Gogh, Vincent. *The Letters*, edited by Leo Jansen, Hans Luijten, and Nienke Bakker. Amsterdam and The Hague: Van Gogh Museum & Huygens ING. Last modified December 2010. http://vangoghletters.org/vg/letters.html

Vaughan, William, Helmut Börsch-Supan, and Hans Joachim Neidhardt. *Caspar David Friedrich, 1774–1840: Romantic Landscape Painting in Dresden*. London: Tate Gallery, 1972.

Vollard, Ambroise. *Recollections of a Picture Dealer*. Translated by Violet M. MacDonald. Boston, MA: Little, Brown & Co., 1936.

Vurro, Milena, Yazhu Ling, and Anya C. Hurlbert. 'Memory Color of Natural Familiar Objects: Effects of Surface Texture and 3-D Shape'. *Journal of Vision* 13, no. 7 (June 2013), 1–20.

Walls, G.L. 'Land! Land!'. *Psychological Bulletin* 57, no. 1 (January 1960), 27–48.

Walsh, Judith. 'A Summer's Pleasure: Incoming Tide, Scarboro, Maine by Winslow Homer'. *The American Art Journal* 25, nos 1–2 (1993), 62–73.

Walsh, Judith. 'Winslow Homer and the Color Theories of Michel-Eugène Chevreul'. In *Martha Tedeschi with Kristi Dahm: Watercolors by Winslow Homer: The Color of Light*, 198–205.

Chicago, IL: Art Institute of Chicago, and New Haven, CT and London: Yale University Press, 2008.

Ward, Martha. *Pissarro, Neo-Impressionism, and the Spaces of the Avant-Garde*. Chicago, IL: The University of Chicago Press, 1996.

Watson, Wendy M. '"Tradition and Innovation": Pierre-Henri Valenciennes and the Neoclassical Landscape'. In *Valenciennes, Daubigny and the Origins of French Landscape Painting*, edited by Michael Marlais, John Varriano, and Wendy Watson, 22–38. State Park, PA: Pennsylvania State University Press, 2004.

Webster, Michael. 'Individual Differences in Color Vision'. In *Handbook of Color Psychology*, edited by A. Elliot, M. Fairchild, and A. Franklin, 197–215. Cambridge: Cambridge University Press, 2015.

Webster, Michael, and Paul Kay. 'Individual and Population Differences in Focal Colors'. In *Anthropology of Color: Interdisciplinary Multilevel Coding*, edited by Robert E. MacLaury, Galina V. Paramei, and Don Dedrick, 29–53. Amsterdam: John Benjamins, 2007.

Webster, Michael, and Deanne Leonard. 'Adaptation and Perceptual Norms in Color Vision'. *Journal of the Optical Society of America A* 25, no. 11 (November 2008), 2817–2825.

Webster, W.R., R.H. Day, and K. Willenberg. 'Orientation-Contingent Color Aftereffects Are Determined by Real Color, Not Induced Color'. *Perception & Psychophysics* 44, no. 1 (January 1988), 43–49.

White, Barbara Ehrlich. *Renoir*. New York: Harry Abrams, 1984.

Wildenstein, Daniel. *Claude Monet: biographie et catalogue raisonné*. 5 vols. Lausanne and Paris: Bibliothèque des Arts, 1974–91.

Willats, John, and Fredo Durand. 'Defining Pictorial Style: Lessons from Linguistics and Computer Graphics'. *Axiomathes* 15, no. 3 (September 2005), 319–351.

Winkler, Alissa D., Lothar Spillmann, John S. Werner, and Michael A. Webster. 'Asymmetries in Blue-Yellow Color Perception and in the Color of "the Dress"'. *Current Biology* 25 (June 2015), R523–R548.

Wittgenstein, Ludwig. *Philosophical Investigations*. Translated by G.E.M. Anscombe. 3rd edn. Oxford: Basil Blackwell, 1995.

Wollheim, Richard. 'On Drawing an Object'. In *On Art and the Mind: Essays and Lectures*, 3–30. London: Allen Lane, 1973.

Wood, Alexander. *Thomas Young: Natural Philosopher 1773–1829*. Cambridge: Cambridge University Press, 2011.

Worrall, David. 'William Blake and Erasmus Darwin's Botanic Garden'. *Bulletin of the New York Public Library* 78, no. 4 (Summer 1975), 397–417.

Wright, Amina. *Joseph Wright of Derby: Bath and Beyond*. Bath: The Holborne Museum, 1988.

Wurmfield, Sanford. 'Color in Abstract Painting'. In *Color for Science, Art and Technology*, edited by Kurt Nassau, 169–194. Amsterdam: Elsevier, 1998.

Yarnall, James L. *John La Farge, a Biographical and Critical Study*. Farnham and Burlington, VT: Ashgate, 2012.

Index

Note: Page numbers in *italics* indicate a figure on the corresponding page.

For Product Safety Concerns and Information please contact our EU representative GPSR@taylorandfrancis.com Taylor & Francis Verlag GmbH, Kaufingerstraße 24, 80331 München, Germany